IB Biology Revision Workbook

IB Biology Revision Workbook

Roxanne Russo

Anthem Press
An imprint of Wimbledon Publishing Company
www.anthempress.com

This edition first published in UK and USA 2020
by ANTHEM PRESS
75–76 Blackfriars Road, London SE1 8HA, UK
or PO Box 9779, London SW19 7ZG, UK
and
244 Madison Ave #116, New York, NY 10016, USA

British Library Cataloguing-in-Publication Data
A catalogue record for this book is available from the British Library.

ISBN-13: 978-1-78527-078-9 (Pbk)
ISBN-10: 1-78527-078-8 (Pbk)

This title is also available as an e-book.

Contents

Preface

A Note to the Student

This guide is for you. It is intended to be used for revision during the two years of the DP Biology course and when studying for your final examinations. The activities will provide you with opportunities to test your knowledge in each of the Standard and Higher Level topics and the Option topic you have studied.

Each activity lists relevant command terms, descriptions of which can be found in the DP Biology guide. The pages and diagrams have been left in black and white for you to add colour to assist in your revision.

Acknowledgements

My students, past and present, for their enthusiasm and shared love of learning biology.

My husband, Rick, for his love and support of everything I do.

General Information

Command terms: list, state and identify

A List of Chemical Compounds and Ions

The table below lists common and important chemical compounds in biology and summarises their function within organisms.

Name	Formula/abbreviation	Example of use/function
Glucose	$C_6H_{12}O_6$	Produced during photosynthesis (Calvin cycle) and broken down during cellular respiration (glycolysis), used as primary energy source in the body
Pyruvate	$C_3H_4O_3$	Product of glycolysis, one glucose splits to form two molecules of pyruvate
Water	H_2O	A useful solvent, used in transport, used in photosynthesis, produced in cellular respiration
Sodium ions	Na^+	Sodium–potassium pumps (active transport)
Potassium ions	K^+	Sodium–potassium pumps (active transport)
Urea	CH_4N_2O	Waste product formed from the breakdown of proteins and passed from the body in urine
Ribose	$C_5H_{10}O_5$	Pentose sugar forming part of the RNA nucleotide
Methane	CH_4	A greenhouse gas
Sodium chloride	$NaCl$	Maintenance of membrane potential, blood volume and blood pressure
Cellulose	$(C_6H_{10}O_5)_n$	Structural component of the plant cell wall
Starch	$(C_6H_{10}O_5)_n$	Energy storage for excess glucose in plants
Glycogen	$C_{24}H_{42}O_{21}$	Energy storage for excess glucose in the liver of animals
Deoxyribonucleic acid	DNA	Composed of nucleotides, contains hereditary information
Ribonucleic acid	RNA	Composed of nucleotides, contains hereditary information, used in transcription (mRNA) and translation (mRNA and tRNA)
Adenosine tri-phosphate	ATP	Energy storage molecule
Chlorophyll	Chlorophyll a or chlorophyll b	Photosynthetic pigment found in chloroplasts
Haemoglobin	Hb or Hgb	Oxygen transport in red blood cells
Carbon dioxide	CO_2	Waste product of cellular respiration, used in photosynthesis

Name	Formula/abbreviation	Example of use/function
Water vapour	H_2O	A greenhouse gas
Nitrogen oxides	N_2O	A greenhouse gas
Acetylcholine	$C_7H_{16}NO_2^+$	A neurotransmitter involved in the peripheral and central nervous systems
Acetyl coenzyme A	Acetyl-CoA	Carries carbon atoms within the acetyl group to the Kreb's cycle in cellular respiration
Nicotinamide adenine dinucleotide	NADH NAD+	Hydrogen carrier. NAD+ accepts electrons and becomes reduced; NADH is the reduced form, which is oxidised and donates electrons. Used in cellular respiration and photosynthesis
Flavin adenine dinucleotide	$FADH_2$ FAD+	Hydrogen carrier. FAD+ accepts electrons and becomes reduced; $FADH_2$ is oxidised and donates electrons. Used in oxidative phosphorylation in cellular respiration
Nicotinamide adenine dinucleotide phosphate	NADP+ NADPH	Hydrogen carrier. NADP+ accepts electrons; NADPH donates electrons. Used in the light dependent reactions (Calvin cycle)
Ribulose bisphosphate	RuBP	A carbon dioxide acceptor involved in the light-independent reactions of photosynthesis
Glycerate-3-phosphate	GP	Formed by the carboxylation of RuBP, then reduced to triose phosphate
Triose phosphate	TP	Converted to glucose, sucrose, starch, fatty acids and amino acids in photosynthesis; regenerates RuBP
Histamine	$C_5H_9N_3$	Involved in the inflammatory response
Calcium ions	Ca^{2+}	Involved in synaptic transmission in neurons and in muscle contraction
Dopamine	$C_8H_{11}NO_2$	A hormone and neurotransmitter that plays a role in the reward system in the brain and in motor control
Serotonin	$C_{10}H_{12}N_2O$	A neurotransmitter involved in feelings of well-being and happiness, regulation of mood, appetite and sleep
Nitrogen	N	Abundant in Earth's atmosphere, component of amino acids and nucleic acids
Ammonia	NH_3	Produced following decay of animal and plant matter; used by the kidneys to neutralise excess acid
Nitrate	NO_3-	Used in fertilisers
Phosphate	$PO_4^{3}-$	Component of nucleic acids, ATP and phospholipids
Ascorbic acid	$C_6H_8O_6$	(Vitamin C) has antioxidant properties, cannot be synthesised by humans
Iron	Fe	Component of haemoglobin and myoglobin
Hydrogen carbonate ions	CHO_3-	(Bicarbonate) involved in the maintenance of the pH level of the blood

Enzymes

The following table lists important enzymes involved in biochemical reactions and summarises their function within cells and organisms.

Enzyme	Source/location	Substrate	Function
Lactase	Small intestine of mammals	Lactose	Breaks down lactose into glucose and galactose
Helicase	Nucleus or nucleoid region	DNA	Unwinds the double helix and separates the two strands by breaking hydrogen bonds during DNA replication
DNA polymerase (SL)	Nucleus	Deoxyribonucleoside triphosphates (dNTPs)	Links nucleotides together to form a new strand
RNA polymerase	Nucleus	DNA, ribonucleoside triphosphates	Separation of DNA strand during transcription, adds the 5' end of the free RNA nucleotide to the 3' end of the growing mRNA molecule
Restriction endonucleases	Nucleus	Specific sequences of double-stranded DNA	Gene transfer to bacteria using plasmids
Amylase	Pancreas	Starch	Digestion of carbohydrates
Lipase	Pancreas	Lipid	Digestion of lipids
Endopeptidases e.g. pepsin, trypsin	Pancreas	Amino acids	Break down peptide bonds between amino acids
DNA polymerase I	Nucleus	RNA primers, deoxyribonucleoside triphosphates (dNTPs)	Removes RNA primer
DNA polymerase III	Nucleus	Deoxyribonucleoside triphosphates (dNTPs)	Adds dNTPs to growing strand at the 3' end of a primer
DNA ligase	Nucleus	Okazaki fragments	Joins Okazaki fragments together, gene transfer to bacteria using plasmids
DNA gyrase	Nucleus	DNA	Relieves the strain
DNA primase	Nucleus	Ribonucleoside triphosphates	Creates RNA primer
tRNA-activating enzymes	Cytoplasm	tRNA, amino acid	Attaches a specific amino acid to a specific tRNA molecule
ATP synthase	Thylakoid membrane of chloroplasts and inner mitochondrial membrane	Adenosine di-phosphate, phosphate	Protons diffuse through ATP synthase to generate ATP
Ribulose bisphosphate carboxylase oxygenase (RuBisCO)	Stroma of chloroplasts	Ribulose bisphosphate, carbon dioxide	Catalyses the carboxylation of ribulose bisphosphate

Synthase – makes something.

Synthetase – uses ATP to make something.

Bonds

Command terms: list, state and identify

Complete the following table to give examples of each type of bond and where it is found or used.

Bond	Description	Examples of where it is found/used
Hydrogen bond	Bonds that hold separate molecules loosely together	
Ionic bond	Bonds that form between ions of opposite charges	
Covalent bond	Strong bonds that occur between non-metal and non-metal	
Peptide bond	Bonds that form between carboxyl and amino groups	
Cross bridge	Bonds that form between muscle filaments	
Di-sulphide bond	A type of covalent bond occurring between two sulphur atoms	

CELL BIOLOGY (TOPIC 1)

Command terms: measure, calculate, estimate, determine and predict

A. Diameter of field of view

The diameter of the field of view is the width of the field of view for a particular magnification on a microscope.

For low power, a ruler with millimetre measurements can be used to actually measure the diameter. For medium- and high-power lenses, the millimetre increments are too large to be seen in the field of view, so a calculation is needed using the measurement taken from the low-power lens.

The formula is:

$$\frac{\text{High-power field of view}}{\text{Low-power field of view}} = \frac{\text{Low-power magnification}}{\text{High-power magnification}}$$

For example:

$$\frac{?}{2,000} = \frac{10 \times 10}{40 \times 10}$$

Therefore, the high-power field of view for this particular microscope:

= 500 μm

The same formula can be used to calculate the field of view for any of the lenses on any light microscope.

B. Size of specimen

The size of the specimen is the actual size viewed under the microscope.

The formula is:

$$\text{Size of specimen} = \frac{\text{Diameter of field of view (μm)}}{\text{No. of times specimen fits across field of view}}$$

For example:

The diagram shows the high-power field of view on a microscope and the cell of interest. The number of times the cell would fit across the field of view is estimated.

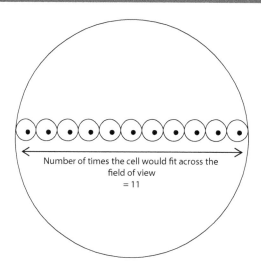

Size of specimen $= \dfrac{500 \ \mu m}{11}$

Therefore, the size of the specimen:

= 45.5 μm

C. Magnification of drawings

The magnification of the drawing is how many times the drawing is larger than the actual size of the specimen. This is essentially the relationship between the size of the actual specimen and the size of the drawing of the specimen.

The formula is:

Magnification of drawing $= \dfrac{\text{Size of drawing of object } (\mu m)}{\text{Size of specimen } (\mu m)}$

For example:

The following diagrams show an electron micrograph of *Dunaliella salina* microalgae (left) and a drawing from the same slide (right).

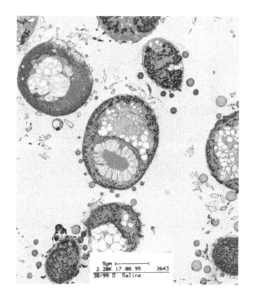

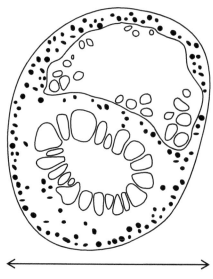

Size of drawing = 6 cm (60,000 μm)

Dunaliella salina.
Courtesy of Adelaide Microscopy, The University of Adelaide, South Australia, used with permission.

Therefore,

$$\text{Magnification of drawing} = \frac{60,000}{20}$$

Therefore, the magnification of this drawing:

= **3000** ×

D. Using scale bars

Scale bars are often used in electron micrographs or diagrams to show size. The scale bar is a line with a measurement above it to show the relationship between the actual length of the line and the distance represented by the line on the drawing.

For example, the electron micrograph below shows that a scale bar of 1 cm in length represents 1 μm on the electron micrograph. This scale can be applied to determine the actual size of any structures within the electron micrograph.

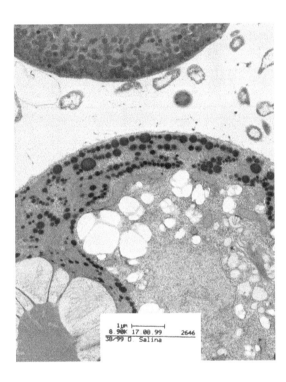

Dunaliella salina.
Courtesy of Adelaide Microscopy, The University of Adelaide, South Australia, used with permission.

1.1 Surface Area to Volume Ratio

Command terms: describe, outline and discuss

Please Note: connections are found throughout the standard level and higher level core and options. Examples shown here are taken from these relevant areas.

Surface area to volume ratio is an essential and underlying concept throughout biology. In order for cells, tissues and organs to function effectively, they require a higher surface area compared with their volume.

The following provides examples of when increased surface area is important; explain *why* this is essential for each example.

Example	Explanation of importance
Movement across membranes by diffusion or osmosis	
Membranes of rough endoplasmic reticulum	
Membranes of Golgi apparatus	
Cellular division by mitosis	
Enzymes and their active sites on substrates	
Skin surface in control of body temperature	
Folding of the inner mitochondrial membrane (cristae)	
Alveoli and pneumocytes in the lungs	
Light harvesting in the photosystems of the chloroplast	
Surface area of leaves	
Chorionic villi	
Villi and microvilli in the small intestine	
Mechanical digestion (chewing and churning)	
Blood vessels (arteries, veins and capillaries)	
Dendrites on neurons	
Supercoiling of DNA	
Thylakoid membranes in chloroplasts	
Root hairs on plant roots	
Fungal hyphae on plant roots	
Capillaries in the Bowman's capsule of the kidney	
Length of the Loop of Henlé	
The placenta	
Folding of the human cerebral cortex	
Sensory hairs on the cochlea	
Bile salts and emulsification of lipids	

1.1 Calculating Magnification

Command terms: measure, calculate, estimate and determine

Below are two electron micrographs and accompanying line drawings. For each, calculate the magnification of the drawing.

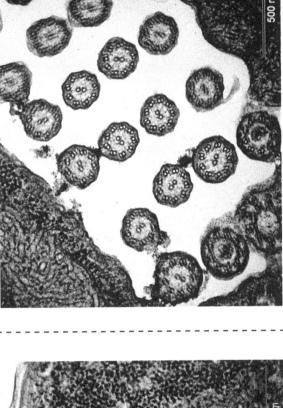

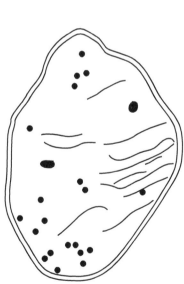

Cilia in Protozoan courtesy of Adelaide Microscopy,
The University of Adelaide,
Australia, used with permission

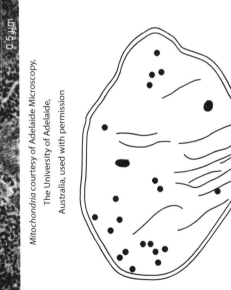

Mitochondria courtesy of Adelaide Microscopy,
The University of Adelaide,
Australia, used with permission

1.2 Eukaryotic Cells

Command terms: draw and label

Colour and label the different parts according to the key provided.

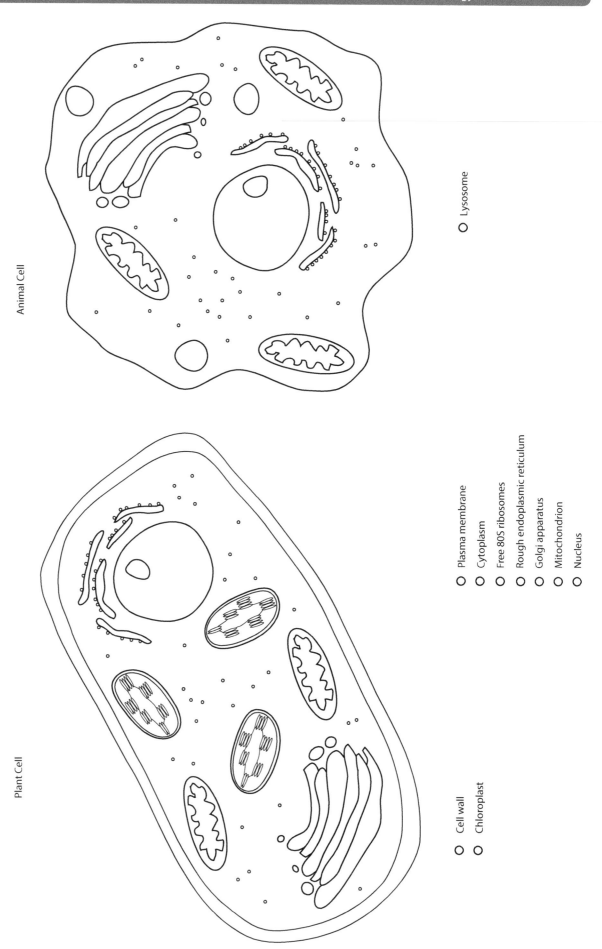

Plant Cell

Animal Cell

○ Cell wall

○ Chloroplast

○ Plasma membrane

○ Cytoplasm

○ Free 80S ribosomes

○ Rough endoplasmic reticulum

○ Golgi apparatus

○ Mitochondrion

○ Nucleus

○ Lysosome

1.2 Plant versus Animal Cells

Command terms: distinguish, compare, compare and contrast

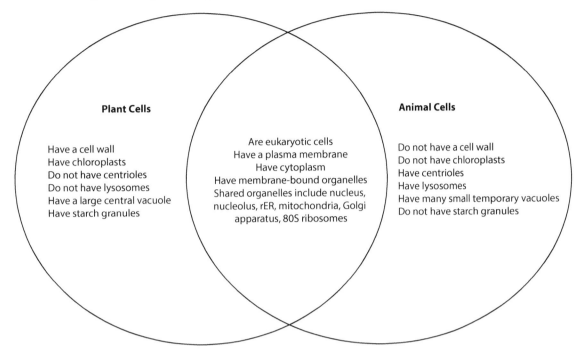

Plant Cells

Have a cell wall
Have chloroplasts
Do not have centrioles
Do not have lysosomes
Have a large central vacuole
Have starch granules

Are eukaryotic cells
Have a plasma membrane
Have cytoplasm
Have membrane-bound organelles
Shared organelles include nucleus, nucleolus, rER, mitochondria, Golgi apparatus, 80S ribosomes

Animal Cells

Do not have a cell wall
Do not have chloroplasts
Have centrioles
Have lysosomes
Have many small temporary vacuoles
Do not have starch granules

1.2 Prokaryotic versus Eukaryotic Cells

Command terms: distinguish, compare, compare and contrast

Prokaryotic Cells

Generally smaller cells
Divide by binary fission
Reproduction is always asexual
Usually unicellular
Circular DNA
Naked DNA (no histones)
DNA located in the cytoplasm in the nucleoid region
Do not have compartmentalisation
Ribosomes are 70 S
May have plasmids
Do not have mitochondria, rER, Golgi apparatus, chloroplasts or lysosomes

Have a plasma membrane
Contain DNA
Have RNA
Have chromosome/s
Have cytoplasm
Have vacuoles
May have flagella
May have cell walls (except animal cells)
Have ribosomes
Have vesicles
Can have extensions of the plasma membrane to aid in transport

Eukaryotic Cells

Generally larger cells
Divide by mitosis or meiosis
Reproduction may be asexual or sexual
Often multicellular
Linear DNA (double helix)
DNA associated with proteins called histones
DNA enclosed in a nucleus
Have a compartmentalised internal structure (organelles are membrane-bound)
Ribosomes are 80 S
Do not have plasmids
Have mitochondria, rER, Golgi apparatus, chloroplasts (plants) and lysosomes (animals)

1.2 Structure and Function of Organelles

Command terms: state and identify

The following organelles are found within either the exocrine gland cells of the pancreas or within palisade mesophyll cells of the leaf. Some of the organelles are present in both.

Match the plant cell and/or animal cell organelles to their function.

Organelle	Function
Nucleus	Contains enzymes, dissolved ions, nutrients and organelles
Plasma membrane	Aerobic cellular respiration
Cytoplasm	Site of protein synthesis for use outside of cell
80S ribosomes	Processing, modification and packaging of proteins
Rough endoplasmic reticulum	Control of cellular activity and cell metabolism
Golgi apparatus	Contains enzymes for breakdown of cellular components
Mitochondrion	Structural support for cells
Cell wall	Site of photosynthesis
Chloroplast	Selective control of entry and exit of materials
Lysosome	Site of protein synthesis for use within the cell

1.2 Prokaryotic Cell

Command terms: draw and label

Colour and label the different parts according to the key provided.

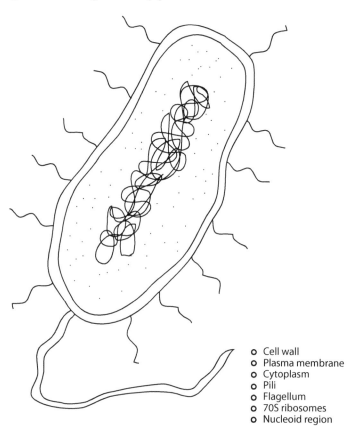

- o Cell wall
- o Plasma membrane
- o Cytoplasm
- o Pili
- o Flagellum
- o 70S ribosomes
- o Nucleoid region

1.2–1.6 Cells – Concept Map

Command terms: define, list, state and identify

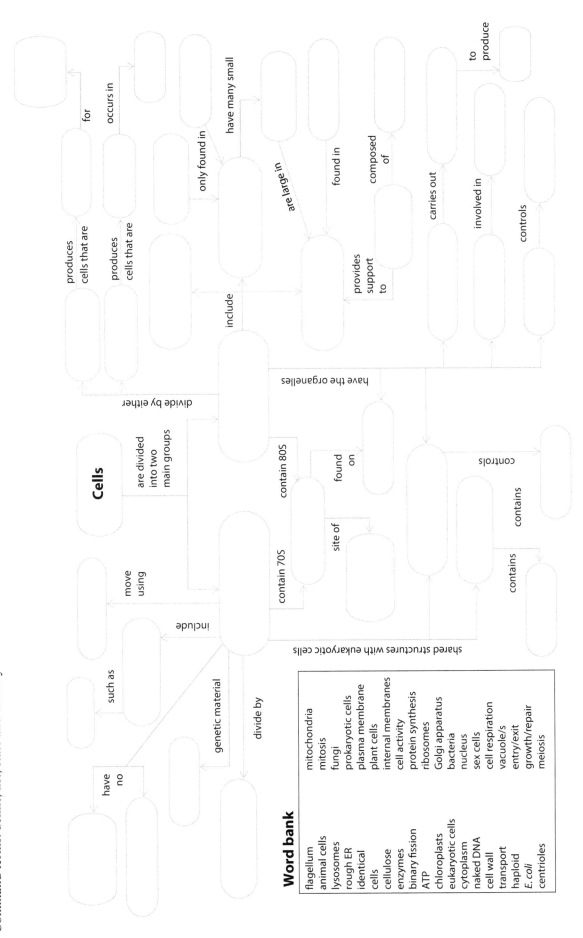

Word bank

flagellum	mitochondria
animal cells	mitosis
lysosomes	fungi
rough ER	prokaryotic cells
identical	plasma membrane
cells	plant cells
cellulose	internal membranes
enzymes	cell activity
binary fission	protein synthesis
ATP	ribosomes
chloroplasts	Golgi apparatus
eukaryotic cells	bacteria
cytoplasm	nucleus
naked DNA	sex cells
cell wall	cell respiration
transport	vacuole/s
haploid	entry/exit
E. coli	growth/repair
centrioles	meiosis

1.3 Fluid Mosaic Model

Command terms: draw, label, state, annotate and identify

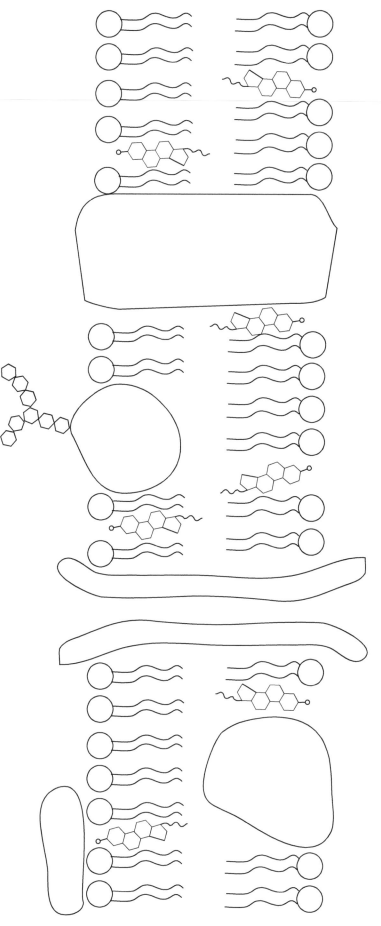

Indicate on the diagram the place/s that the following processes occur:

● osmosis   ■ simple diffusion

◀ facilitated diffusion   ◆ active transport

★ endocytosis / exocytosis

Label the parts listed below. Colour each part a different colour and fill out the key.

○ phospholipid molecule   ○ hydrophilic head   ○ glycoprotein   ○ integral protein

○ cholesterol   ○ hydrophobic tail   ○ channel protein

                           ○ transmembrane protein ○ peripheral protein

1.3, 1.4 Cell Membranes – Concept Map

Command terms: define, list, state and identify

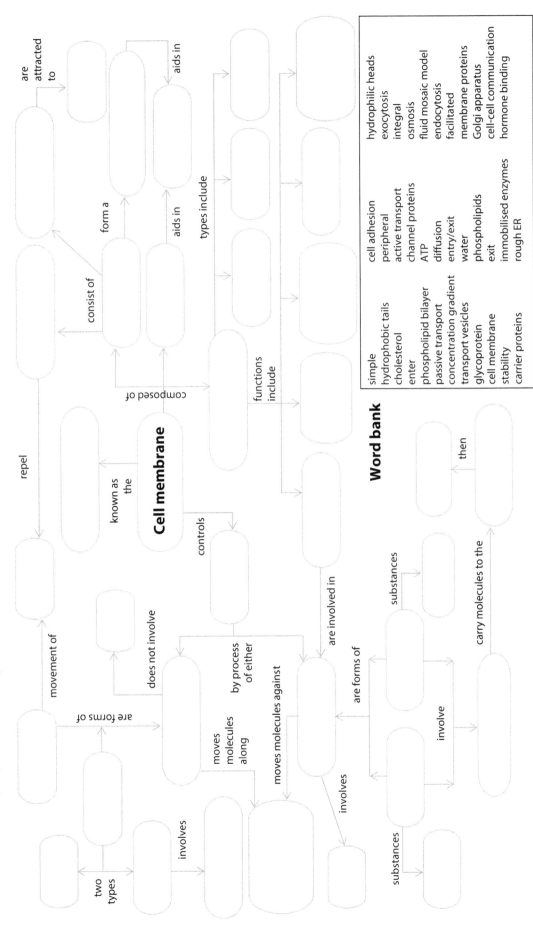

Word bank

simple	hydrophilic heads
hydrophobic tails	exocytosis
cholesterol	integral
enter	osmosis
phospholipid bilayer	fluid mosaic model
passive transport	endocytosis
concentration gradient	facilitated
transport vesicles	membrane proteins
glycoprotein	Golgi apparatus
cell membrane	cell-cell communication
stability	hormone binding
carrier proteins	
	cell adhesion
	peripheral
	active transport
	channel proteins
	ATP
	diffusion
	entry/exit
	water
	phospholipids
	exit
	immobilised enzymes
	rough ER

1.4 Transport across Membranes

Command terms: distinguish, compare, compare and contrast

Complete the table to summarise the methods of transportation across cell membranes.

	Active or passive?	ATP	Concentration gradient	Proteins involved?	Example
Osmosis					
Simple diffusion					
Facilitated diffusion					
Active transport (protein pumps)					
Endocytosis					
Exocytosis					

1.4 Endocytosis versus Exocytosis

Command terms: label, annotate and identify

The diagrams below show the events occurring during the entry of material to a cell by endocytosis and exit of materials from the cell by exocytosis.

1. Label the parts shown in the two diagrams.
2. Draw arrows on the diagram to show the direction of movement of the molecules.
3. Annotate the arrows you have drawn to outline the events of both processes.

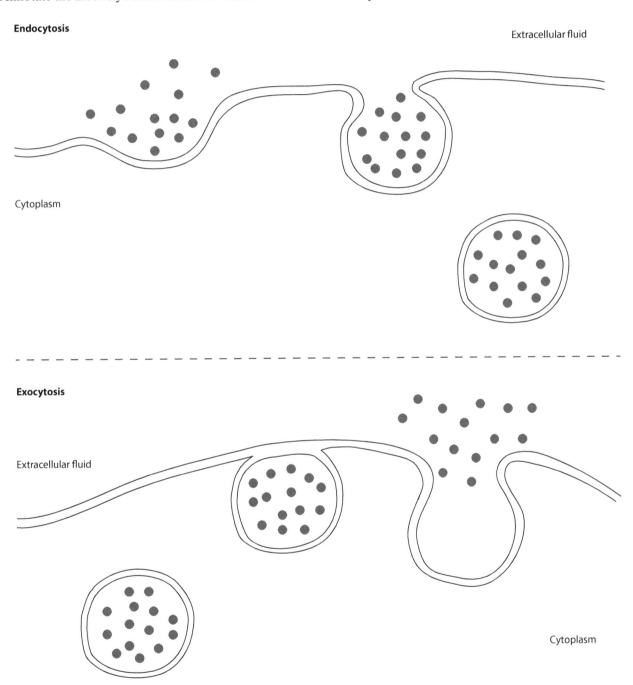

Endocytosis

Extracellular fluid

Cytoplasm

Exocytosis

Extracellular fluid

Cytoplasm

1.5 Endosymbiotic Theory

Command terms: state, describe, outline and explain

Fill in the blanks to complete the sentences below that outline the evidence for the endosymbiotic theory.

Organelles evolved from independent prokaryotes that were _____ by larger cells by _____.

Eukaryotic cells contain _____ and _____, neither of which are found in _____ cells.

These smaller cells survived inside the larger cells in a _____ relationship.

The smaller cells continued to carry out the processes of _____ and _____.

These smaller cells are thought to have developed into the organelles _____ and _____ because of the characteristics similar to prokaryotic cells.

Both organelles have _____ DNA and _____ ribosomes-like prokaryotes.

Both organelles have a _____ membrane because they were taken in to _____ by endocytosis.

1.4 Membrane Transport

Command terms: annotate, describe, outline and explain

1. Add arrows to show the direction of movement for each transport type.
2. Annotate the diagram to show the type of molecule moving for each transport type.
3. Identify those types of transport that require ATP.

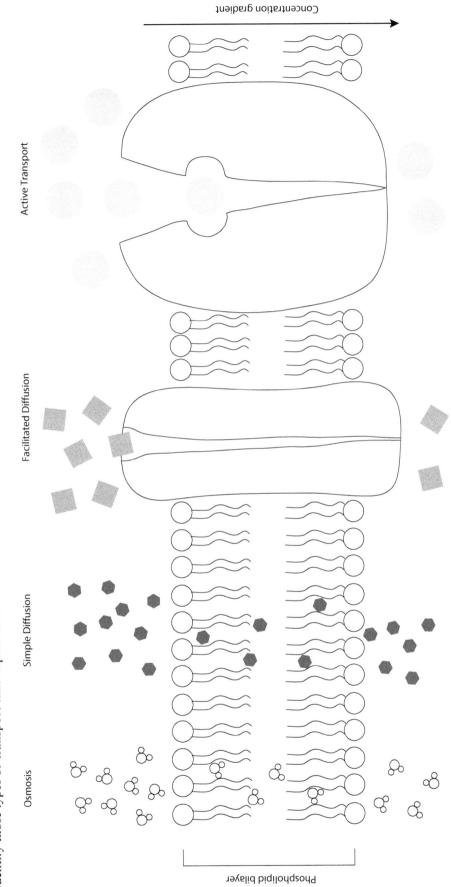

1.6　Cytokinesis

Command terms: distinguish, compare, compare and contrast

Complete the table to summarise the differences in cytokinesis between plant and animal cells.

	Animal cell	Plant cell
Cell plate present		
Contractile ring		
Cleavage furrow		
Number of daughter cells		
Involvement of vesicles		

1.6　Determining Mitotic Index

Command terms: measure, calculate, estimate, determine and predict
1. Obtain a prepared slide or make your own slide of the root tip of an onion.
2. Focus on high power and find a region on the slide where there are many cells undergoing cell division.
3. Create a table to tally your results as follows:

Stage of mitosis	Number of cells in each stage
Interphase[a]	
Prophase	
Metaphase	
Anaphase	
Telophase	

[a] Not technically a stage of mitosis.

4. Classify each of around 100 cells either as being in any of the stages of mitosis or as being in interphase.
5. Use the data collected in your table to calculate the mitotic index, which is determined using the following equation.

Mitotic index = number of cells in mitosis/total number of cells

  ○ The mitotic index can be an important tool to determine the presence of tumours and categorise them.

1.6 Mitosis

Command terms: identify, deduce and determine

Categorise the following events into the correct phase of mitosis; either prophase, metaphase, anaphase or telophase.

Centromeres divide _____

Spindle microtubules disappear _____

Chromosomes are visible _____

Chromosomes pulled to opposite poles by microtubules _____

Nuclear membrane reforms _____

Chromatids now known as chromosomes _____

Chromosomes decondense _____

Nuclear membrane is completely broken down _____

Cell plate forms in plant cells only _____

Chromosomes line up along equator _____

Chromosomes are visible _____

Spindle microtubules grow between poles and equator _____

Spindle microtubules attach to centromeres _____

Chromosomes condense and supercoil _____

Chromosomes separate into two chromatids _____

Nuclear membrane breaks down _____

Chromatids have fully separated due to breaking of the centromere _____

Chromosomes consist of two identical sister chromatids _____

Nucleolus disappears _____

1.6 Mitosis

Command terms: draw, label, annotate and sketch

Complete the boxes to sketch each stage of mitosis and include annotations to explain the movement of the chromosomes during that stage.

Prophase	Description

Metaphase	Description

Anaphase	Description

Telophase	Description

1.6, 3.3 Cell Division – Concept Map

Command terms: define, list, state and identify

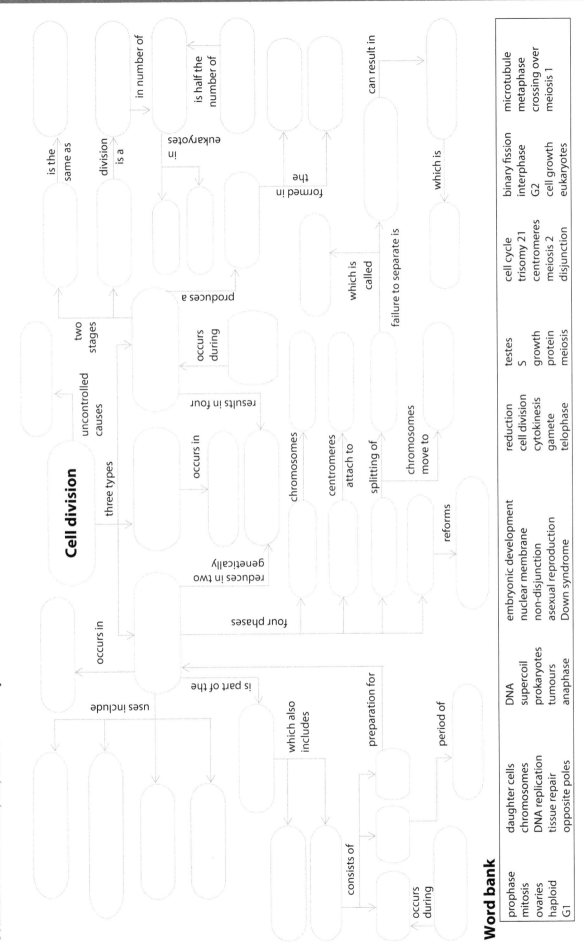

Word bank

prophase	DNA	embryonic development	reduction
mitosis	supercoil	nuclear membrane	cell division
ovaries	prokaryotes	non-disjunction	cytokinesis
haploid	tumours	asexual reproduction	gamete
G1	anaphase	Down syndrome	telophase

testes	binary fission	microtubule
S	interphase	metaphase
growth	G2	crossing over
protein	cell growth	meiosis 1
meiosis	eukaryotes	

cell cycle	
trisomy 21	
centromeres	
meiosis 2	
disjunction	

Labels within the concept map: is the same as; in number of; division is a; is half the number of; in eukaryotes; formed in the; can result in; which is; two stages; produces a; occurs during; which is called; failure to separate is; uncontrolled causes; results in four; occurs in; chromosomes; centromeres attach to; splitting of; chromosomes move to; Cell division; three types; reduces in two genetically; four phases; reforms; occurs in; uses include; is part of the; which also includes; preparation for; period of; consists of; occurs during

1.6 Smoking and Cancer

Command terms: state, calculate, describe, estimate, identify, deduce, evaluate and suggest

The following graph shows the incidence of cancer and the mortality rate in male smokers. The questions below refer to this graph.

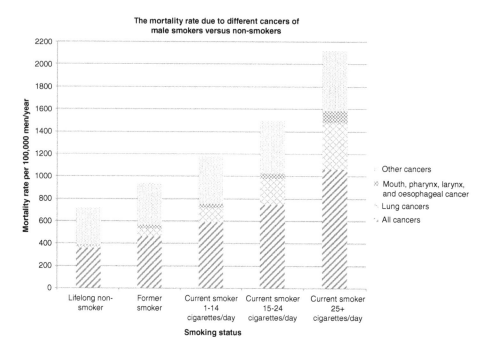

1. Calculate the percentage difference in mortality rate of lung cancer in men who smoke 1–14 cigarettes per day to men who smoke 25+ cigarettes per day.

   _____

   _____

2. Compare the effect of the number of cigarettes smoked per day on the mortality rate of mouth, pharynx, larynx and oesophageal cancers.

   _____

   _____

   _____

3. The correlations shown in the graph do not necessarily provide evidence of causation. Discuss this in relation to cancer and smoking.

   _____

   _____

   _____

4. Explain how cigarette smoke, as an example of a carcinogen, causes cancer in the body. Reference should be made to the following terms: mutagen, oncogene, metastasis, primary tumour, secondary tumour.

   _____

   _____

   _____

   _____

   _____

   _____

MOLECULAR BIOLOGY (TOPIC 2)

2.1, 2.3, 2.4, 2.6 Features of Macromolecules

Command terms: identify, deduce and determine

Classify the following as features of either (a) carbohydrates, (b) lipids, (c) proteins or (d) nucleic acids.

Long-term energy storage   _____

Monomers are monosaccharides   _____

Short-term energy storage   _____

Allow for buoyancy   _____

Contain nitrogen   _____

May be saturated or unsaturated   _____

Contain peptide linkages   _____

Stored as glycogen in animal livers   _____

Have four levels of structure   _____

Function as thermal insulation   _____

Building blocks are nucleotides   _____

Contain carboxylic acids   _____

Produced following photosynthesis   _____

Function as enzymes   _____

Many have names ending in –ose   _____

May be denatured   _____

Stored as oils in plants   _____

Synthesised at ribosomes   _____

Builds the genetic code   _____

Store high-energy content per gram   _____

Many have names ending in –ase   _____

Broken down during cell respiration   _____

Structural part of plasma membranes   _____

2.1 Molecular Diagrams

Command terms: identify, deduce and determine

Use the names of the molecules given below to correctly identify the molecular diagrams.

Alpha-D-glucose   Beta-D-glucose   D-ribose   Amino acid

Deoxyribose   Saturated fatty acid

2.2 Properties of Water

Command terms: list, state and outline

Complete the following table to summarise the properties of water and the benefit of these properties to living organisms.

Property	Explanation	Example of benefit to living organisms
Cohesion		
	The attraction between molecules of different types	
Thermal		
	As water is polar, many substances are able to dissolve in water	

2.3 Determination of Body Mass Index

Command terms: measure, calculate, estimate, determine and predict

Body Mass Index (BMI) can be used to measure if a person's body mass is within a healthy level. The two methods of determining BMI are with a formula or a special chart called a nomogram.

There are some problems with the use of BMI as the only measure of a person's health. These are listed as follows:
- BMI does not distinguish between genders, males and females store fat differently
- BMI does not distinguish fat from muscle or water retention, so body mass is often not an indication of body fat
- BMI does not take into account race or ethnicity
- BMI is only useful for adults aged over 18
- BMI cannot be used for pregnant women

The table below shows the BMI and associated weight status.

BMI	Status
Below 18.5	Underweight
18.5–24.9	Normal
25.0–29.9	Overweight
30.0 and over	Obese

1. Using the formula:

$$BMI = \frac{Weight\ (kg)}{Height\ (m)^2}$$

- Measure the person's weight in kilograms and height in metres
- Divide the weight by the height, then divide the answer by the height again

For example:
  Rick's height is 1.88 m and weight is 89 kg.
  89/1.88 = 47.34
  47.34/1.88 = 25.18

2. Using a nomogram:

Find the body weight on the right axis and the height on the left axis. Draw a line connecting these two points. The two points where they meet on the BMI scale in the centre is the BMI.

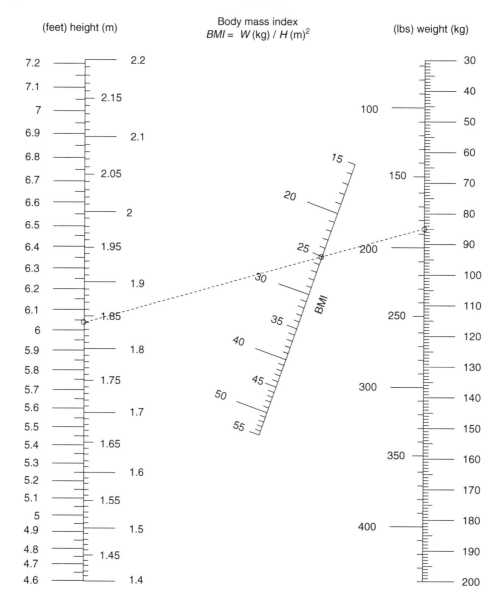

For example:

In the nomogram above, the person's height is 1.84 m and weight is 95 kg. The person's BMI is 25.

2.3–2.4 Condensation and Hydrolysis

Command terms: state, distinguish and identify

Complete the following reactions and categorise them as either (a) condensation or (b) hydrolysis

Carbohydrates:

Glucose + _____ → maltose + _____

Sucrose + _____ → glucose + _____

Lipids:

Water + monoglyceride → _____ + _____

_____ + 2 fatty acids → _____ + water

Proteins:

Amino acid + _____ → _____ + _____

Polypeptide + _____ → _____ + _____

2.5 Enzyme Activity

Command terms: draw, label, annotate, outline, construct and sketch

Complete the following graphs to show the effect of each factor on enzyme activity. Annotate each graph to explain reasons for its shape.

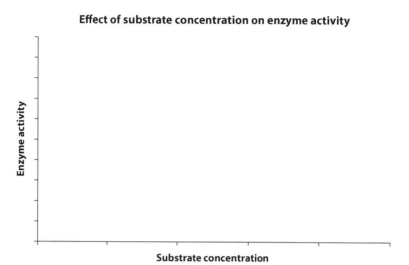

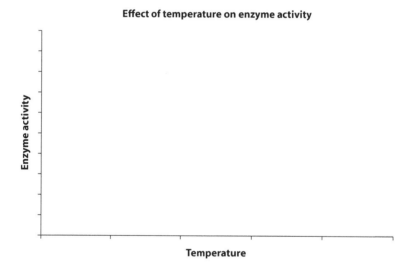

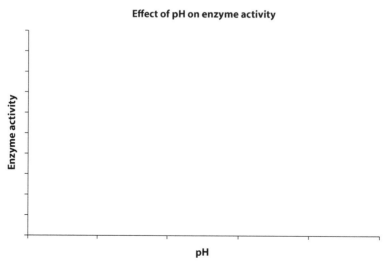

2.6 DNA – Concept Map

Command terms: define, list, state and identify

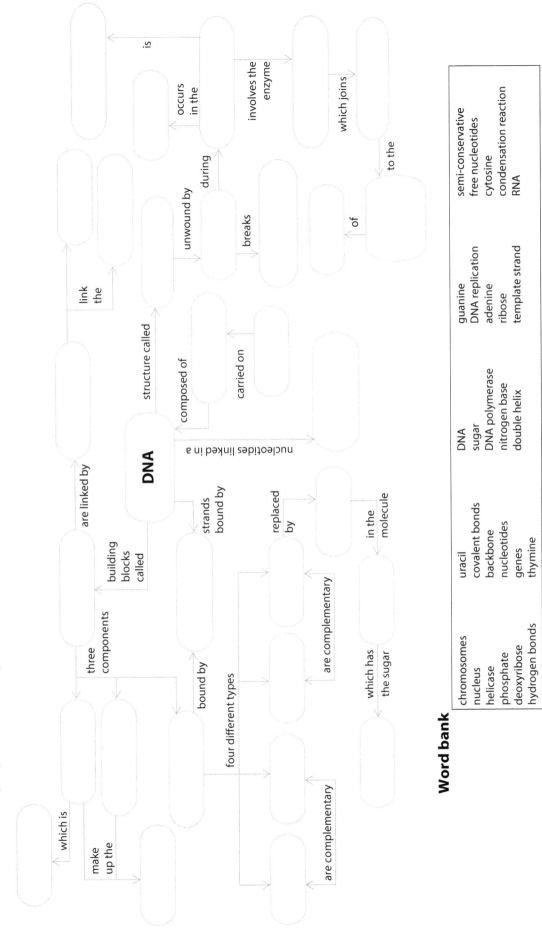

Word bank

chromosomes	uracil	DNA	guanine	semi-conservative
nucleus	covalent bonds	sugar	DNA replication	free nucleotides
helicase	backbone	DNA polymerase	adenine	cytosine
phosphate	nucleotides	nitrogen base	ribose	condensation reaction
deoxyribose	genes	double helix	template strand	RNA
hydrogen bonds	thymine			

2.6 RNA versus DNA

Command terms: distinguish, compare, compare and contrast

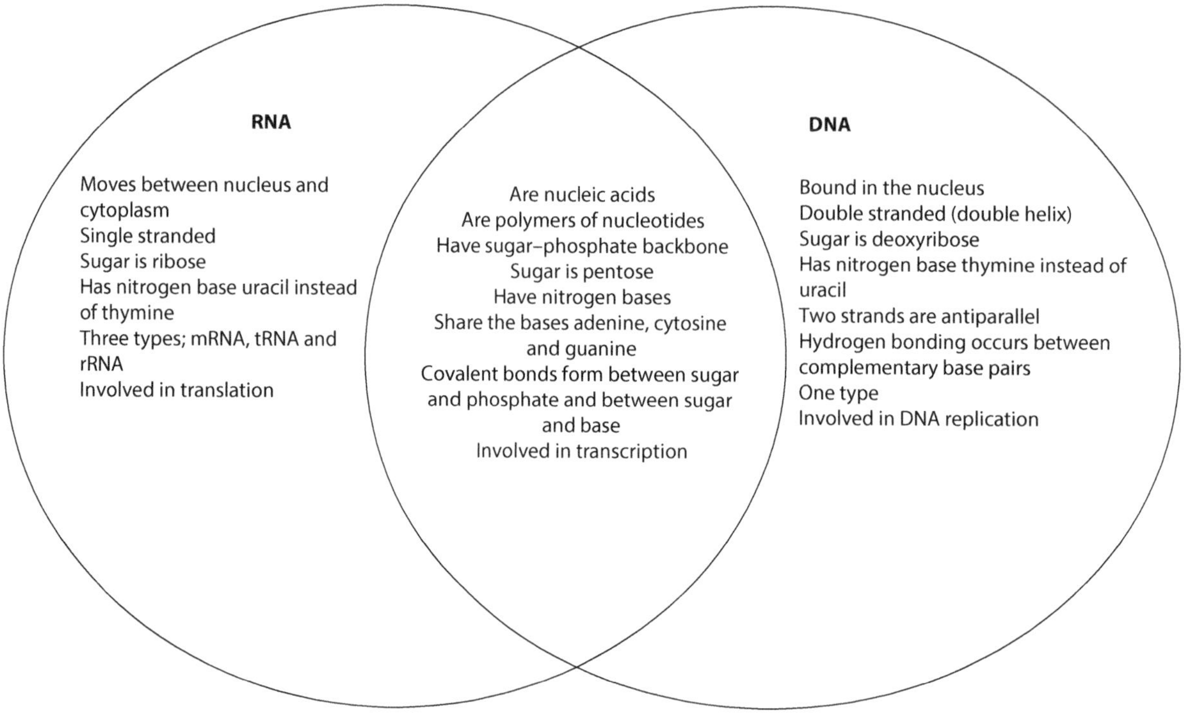

RNA

Moves between nucleus and cytoplasm
Single stranded
Sugar is ribose
Has nitrogen base uracil instead of thymine
Three types; mRNA, tRNA and rRNA
Involved in translation

Are nucleic acids
Are polymers of nucleotides
Have sugar–phosphate backbone
Sugar is pentose
Have nitrogen bases
Share the bases adenine, cytosine and guanine
Covalent bonds form between sugar and phosphate and between sugar and base
Involved in transcription

DNA

Bound in the nucleus
Double stranded (double helix)
Sugar is deoxyribose
Has nitrogen base thymine instead of uracil
Two strands are antiparallel
Hydrogen bonding occurs between complementary base pairs
One type
Involved in DNA replication

2.7 DNA Replication versus Protein Synthesis

Command terms: distinguish, compare, compare and contrast

The following table compares DNA replication with protein synthesis (transcription and translation).

	DNA replication	Protein synthesis
Location	Nucleus	Nucleus and ribosomes
Enzyme that unzips DNA	Helicase	RNA polymerase
Strand separation	Helicase	RNA polymerase
Type of new nucleotides added	DNA	RNA
Addition of new nucleotides	DNA polymerase	RNA polymerase
Nucleic acids involved	DNA	DNA, mRNA, tRNA
Direction	5'–3'	5'–3'
Strand formed	Double	Single
Purpose	Increase the amount of DNA before cell division	Use genetic code from DNA to make polypeptides
Final product/s	Semi-conservative DNA	Polypeptides

2.3, 2.4, 2.6 Macromolecule Summary Chart

Command terms: distinguish, compare, compare and contrast

Complete the following table to summarise the properties of each of the macromolecules.

Macromolecule	Elements	Building block/monomer	Function	Examples
Carbohydrate			Structural	
	CHO	Fatty acids and glycerol	Structural	
Protein	CHON			Enzymes
		Nucleotide		DNA, RNA

2.7 Transcription versus Translation

Command terms: distinguish, compare, compare and contrast

Complete the following table to compare the two stages of protein synthesis.

	Transcription	Translation
Location		
Enzymes involved		
Nucleic acids involved		
Direction		
Final product/s		

2.7 Protein Synthesis – Concept Map

Command terms: define, list, state and identify

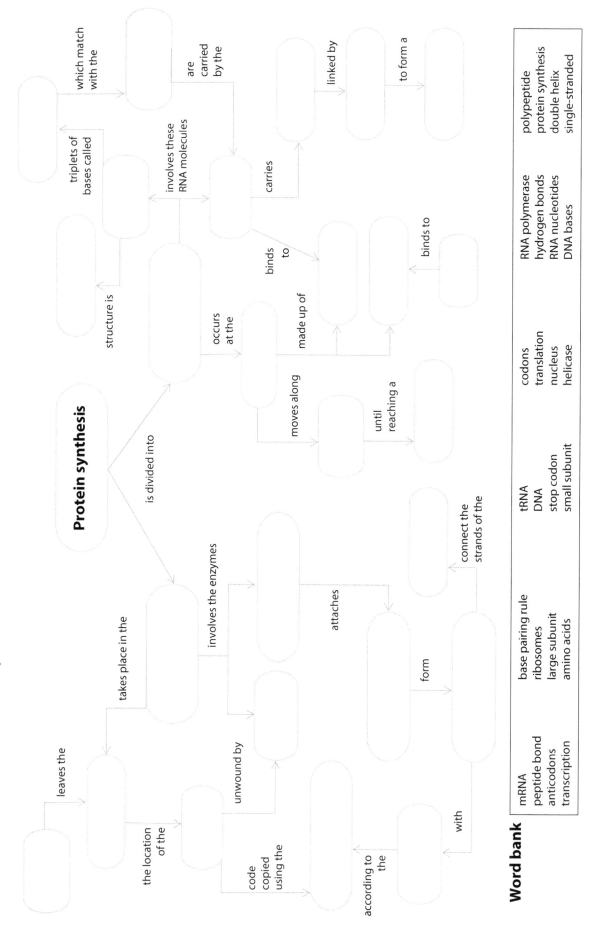

Word bank

mRNA	base pairing rule	tRNA	codons	polypeptide
peptide bond	ribosomes	DNA	translation	protein synthesis
anticodons	large subunit	stop codon	nucleus	double helix
transcription	amino acids	small subunit	helicase	single-stranded
			RNA polymerase	
			hydrogen bonds	
			RNA nucleotides	
			DNA bases	

2.7 Protein Synthesis

Command terms: outline, describe and explain

Number the following statements in the correct order to show the process of protein synthesis.

tRNA brings the corresponding amino acid to the ribosome   _____

tRNA has a complementary anticodon to the codon on the mRNA   _____

tRNA matches its anticodon to the corresponding codon on the mRNA and   _____
hydrogen bonds are formed

Dipeptide bonds are formed to link the two amino acids together   _____

The mRNA brings the code to the ribosome and binds to the small subunit   _____

mRNA carries codons formed by three bases   _____

mRNA copies the code from the template strand of the DNA using   _____
complementary base pairing

The DNA is unwound by RNA polymerase, exposing the bases   _____

The DNA is rewound by RNA polymerase   _____

Free RNA nucleotides assemble along the template strand of DNA   _____

RNA nucleotides join to form a strand of mRNA   _____

mRNA leaves the nucleus via a nuclear pore   _____

The ribosome moves along the mRNA strand   _____

A stop codon is reached   _____

The polypeptide is released   _____

2.7 Amino Acid Codon Table

Command terms: list, state and identify

The following table summarises each of the twenty essential amino acids and their corresponding mRNA codons.

Amino acid	Amino acid 3-letter code	Polar/non-polar	mRNA codon/s
Alanine	Ala	Non-polar	GCU, GCC, GCA, GCG
Arginine	Arg	Polar (+)	CGU, CGC, CGA, CGG, AGA, AGG
Asparagine	Asn	Polar	AAU, AAC
Aspartic acid	Asp	Polar (−)	GAU, GAC
Cysteine	Cys	Polar	UGU, UGC
Glutamic acid	Glu	Polar (−)	GAA, GAG
Glutamine	Gln	Polar	CAA, CAG
Glycine	Gly	Non-polar	GGU, GGC, GGA, GGG
Histidine	His	Polar (+)	CAU, CAC

Amino acid	Amino acid 3-letter code	Polar/non-polar	mRNA codon/s
Isoleucine	Ile	Non-polar	AUU, AUC, AUA
Leucine	Leu	Non-polar	CUU, CUC, CUA, CUG, UUA, UUG
Lysine	Lys	Polar (+)	AAA, AAG
Methionine	Met	Non-polar	AUG
Phenylalanine	Phe	Non-polar	UUU, UUC
Proline	Pro	Non-polar	CCT, CCC, CCA, CCG
Serine	Ser	Polar	UCU, UCC, UCA, UCG, AGU, AGC
Threonine	Thr	Polar	ACU, ACC, ACA, ACG
Tryptophan	Trp	Non-polar	UGG
Tyrosine	Tyr	Polar	UAU, UAC
Valine	Val	Non-polar	GUU, GUC, GUA, GUG
Stop codons			UAA, UAG, UGA

2.8 Anaerobic versus Aerobic Respiration

Command terms: distinguish, compare, compare and contrast

Complete the following table to compare the process of anaerobic respiration with the process of aerobic respiration.

	Anaerobic respiration	Aerobic respiration
Oxygen required?		
Glucose required?		
Glycolysis used?		
ATP production		
Pyruvate production		
CO_2 production		
Final products in animals		
Final products in plants/ yeast		
Hydrogen carriers		
Location		

2.9 Photosynthesis – Concept Map

Command terms: define, list, state and identify

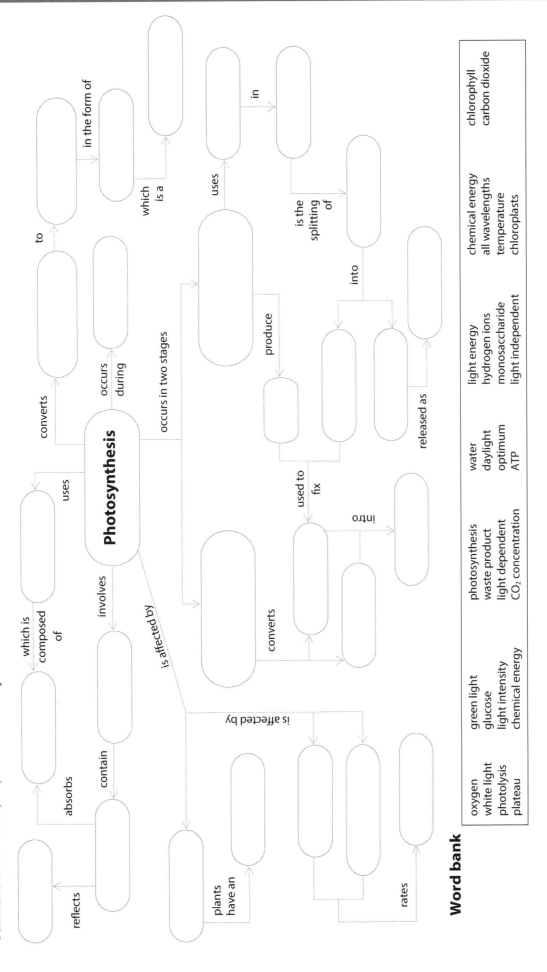

Word bank

oxygen	green light	photosynthesis	water	light energy	chlorophyll
white light	glucose	waste product	daylight	hydrogen ions	carbon dioxide
photolysis	light intensity	light dependent	optimum	monosaccharide	chemical energy
plateau	chemical energy	CO_2 concentration	ATP	light independent	all wavelengths
					temperature
					chloroplasts

2.9 Action Spectrum and Absorption Spectrum

Command terms: draw, label, annotate, construct and sketch

Draw the absorption spectrum for chlorophyll and the action spectrum for photosynthesis on the axes below.

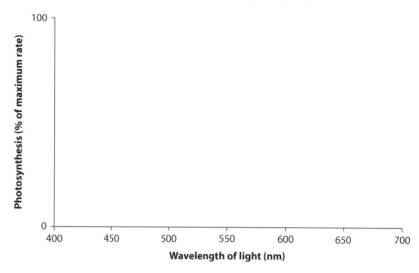

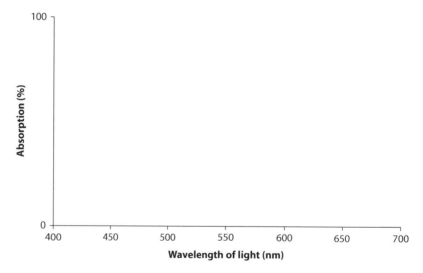

GENETICS (TOPIC 3)

3.1 Sickle Cell Anaemia

Command terms: state, describe, outline and explain

Fill in the blanks to complete the sentences below that outline the characteristics of the genetic disease, sickle cell anaemia.

Sickle cell anaemia affects chromosome number _____.

It is caused by the smallest possible mutation called _____.

The base sequence of _____ is mutated to _____ on the _____ codon.

This means that a _____ with a different anticodon attaches.

In this mutation, the amino acid _____ is replaced with _____.

As a result of this mutation, a different _____ is synthesised, in this case, a distorted _____ _____ molecule.

Sickle cell anaemia affects the shape of _____ reducing its ability to carry _____.

The sickled cells also cause damage to tissues by blocking _____ and reducing _____.

The life of the red blood cells is also reduced from _____ to as little as _____ days.

3.1 Number of Genes

Command terms: compare and state

The table below summarises the number of genes found in a range of different species.

Species name	Common name	Number of genes
Homo sapiens	Human	23,000
Canis familiaris	Dog	25,000
Oryza sativa	Rice	41,000
Saccharomyces cerevisiae	Yeast	6,000
Drosophila melanogaster	Fruit fly	14,000
Escherichia coli	E. coli	3,200

3.2 Comparison of Genome Size

Command terms: compare and state

The genome size is the total length of DNA in an organism. The table below summarises the genome size in millions of base pairs for a range of different species.

Species name	Common name/description	Genome size (million base pairs)
T2 phage	An enterobacteria phage (virus)	0.18
Escherichia coli	E. coli (bacteria)	5
Drosophila melanogaster	Fruit fly	140
Homo sapiens	Human	3,000
Paris japonica	Japanese canopy plant	150,000

3.2 Comparison of Diploid Chromosome Numbers

Command terms: compare and state

The table below summarises the diploid number of chromosomes present in a number of different species.

Species name	Common name	Diploid chromosome number
Homo sapiens	Human	46
Pan troglodytes	Chimpanzee	48
Canis familiaris	Dog	78
Oryza sativa	Rice	24
Parascaris equorum	Horse roundworm	4

3.2 Prokaryotic and Eukaryotic Chromosomes

Command terms: distinguish, compare, compare and contrast

Complete the following table to summarise the similarities and differences in the chromosomes found in prokaryotes and eukaryotes.

	Prokaryotes	Eukaryotes
Number of chromosomes		
Location		
Shape		
Pairing?		
Chromosomes made of		
Introns present?		
Plasmids present?		
Histones present?		
Different chromosomes carry different genes?		

3.3 Stages of Meiosis

Command terms: state, identify, annotate and outline

The diagrams below and on the following page show the various stages of meiosis I and II.

Identify the stage shown in each of the diagrams and annotate the diagram with an outline of the events occurring during each stage.

Number the diagrams to show the correct order of the stages. Prophase I has been completed for you.

1 Name of stage

Prophase I

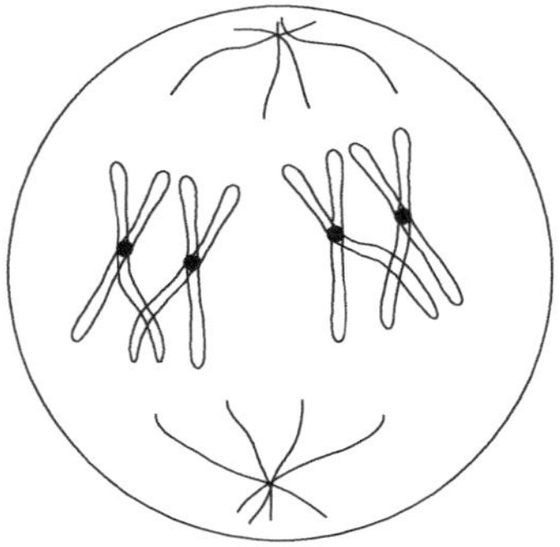

○ Name of stage

○ Name of stage

○ Name of stage

Outline of events

Cell has 2n chromosomes

Chromosomes condense

Spindle forms

Homologous chromosomes

pair

Crossing over occurs

_____

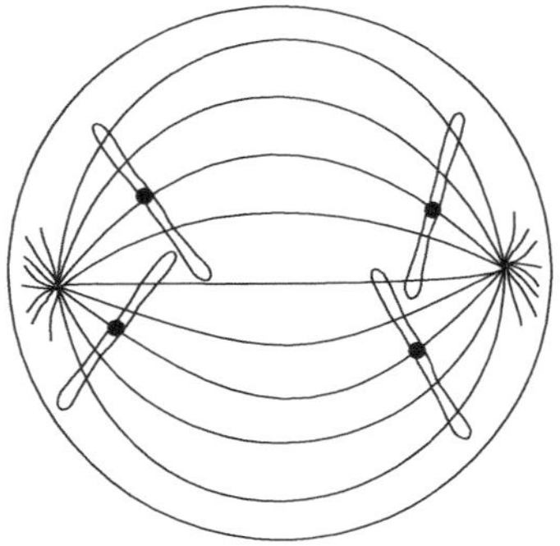

Outline of events

_____
_____
_____
_____
_____

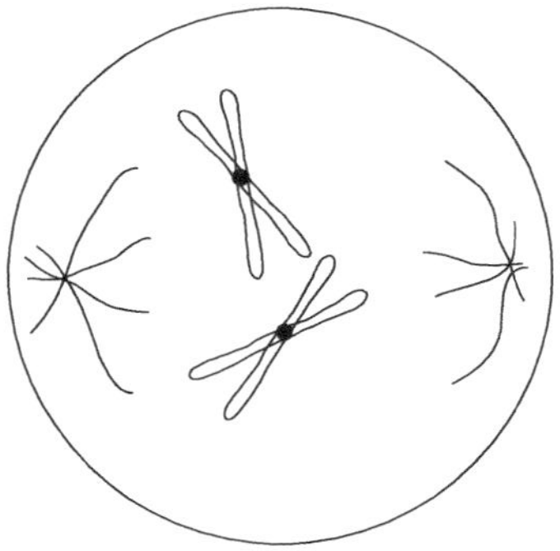

Outline of events

_____
_____
_____
_____
_____

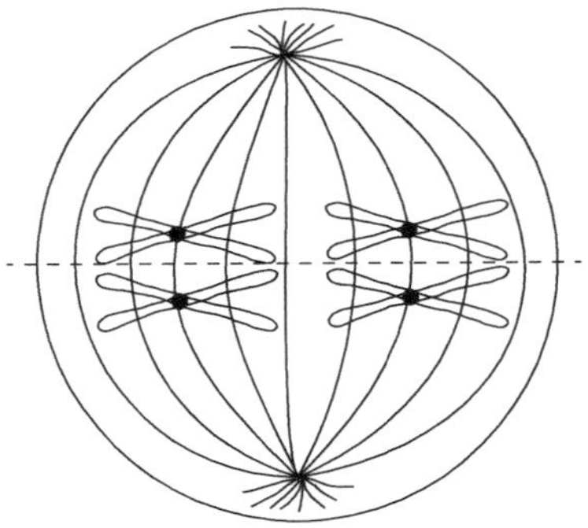

Outline of events

_____
_____
_____
_____
_____

3.3 Stages of Meiosis

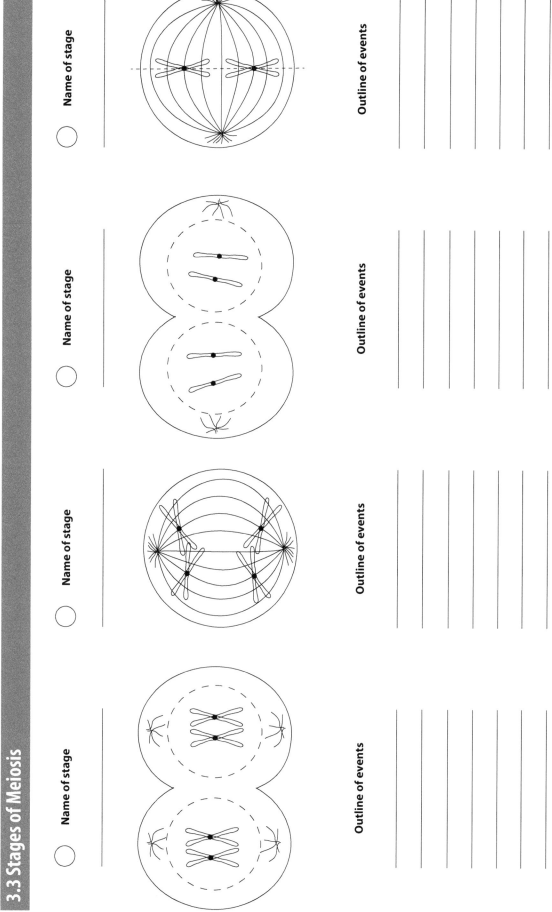

Name of stage

Outline of events

Name of stage

Outline of events

Name of stage

Outline of events

Name of stage

Outline of events

Mitosis versus Meiosis

Mitosis

Produces diploid cells
Two daughter cells produced
Genetically identical cells produced
Occurs in all somatic cells
Used for growth, tissue repair, embryonic development and asexual reproduction
One division
Number of chromosomes remains the same
Chromatids separate in Anaphase

Nuclear division
Involve stages prophase, metaphase, anaphase and telophase
Shared genetic material between parent and daughter cells
Chromosomes and DNA replicate before division
Each new daughter cell has the same number of chromosomes as the other daughter cell/s
Nuclear membrane breaks down
Chromosomes line up on equator during metaphase
Utilise spindle fibres
Interphase (before) and cytokinesis (after) occur

Meiosis

Produces haploid cells
Four daughter cells produced
Genetically different cells produced
Results in increased variation
Occurs in the sex cells (ovaries and testes)
Produces gametes for sexual reproduction
Two divisions; meiosis I and meiosis II
Number of chromosomes is halved
Synapsis and crossing over occur in prophase I
Homologous chromosomes pair during prophase I
Independent assortment occurs in metaphase I
Homologous chromosome pairs separate in anaphase I

3.4 Inheritance of ABO Blood Groups

Command terms: identify, state and describe

Fill in the blanks to complete the sentences below that describe the inheritance of the ABO blood groups in humans.

The ABO blood system in humans is an example of both _____ and _____

There are three _____ involved in this system.

The alleles of _____ are said to be co-dominant.

The allele for blood group o is _____

There are _____ different blood groups possible.

Two of the blood groups have two possible _____

In order to develop the _____ of blood group AB, the genotype must be _____

This results, as the alleles of _____ do not mask one another.

To develop blood group o, the genotype must be _____

To develop blood group A or blood group B, the genotype may be _____ or _____
_____

The possible genotypes for blood group A are _____ and for blood group B are _____

Complete the following table showing the phenotypes and genotypes of each ABO blood group:

Phenotype	Genotype/s
A	
B	
AB	
O	

3.4 Analysis of Pedigree Charts

Command terms: identify, analyse, deduce, predict and suggest

Consider the following pedigree chart, which shows the inheritance of red–green colour blindness in a family.

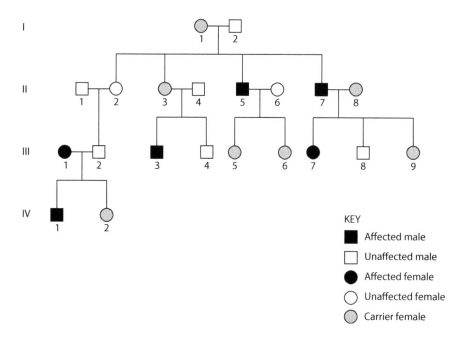

KEY

■ Affected male

□ Unaffected male

● Affected female

○ Unaffected female

◐ Carrier female

Is red–green colour blindness a sex-linked or autosomal condition?

Explain using examples from the chart.

_____

_____

_____

Is red–green colour blindness dominant or recessive?

Explain using examples from the chart.

_____

_____

_____

Why are there no male carriers shown in the pedigree chart?

_____

_____

_____

3.5 Genetic Modification and Biotechnology

Command term: state

Complete the following table to state the purpose and product and describe the process of the techniques listed.

Technique	Purpose	Description of process involved	Product
Gel electrophoresis			
Polymerase chain reaction			
DNA profiling			
Gene transfer			
Cloning			

ECOLOGY (TOPIC 4)

4.1 Species, Communities and Ecosystems

Command terms: define and state

Link the following terms with their meanings.

Species	An organism that obtains organic nutrients from dead organisms by external digestion
Autotroph	A group of organisms of the same species living and interacting in the same place at the same time
Heterotroph	A combination of a community and the abiotic environment
Detritivore	A group of organisms that can interbreed to produce fertile offspring
Saprotroph	Populations of different species living and interacting together
Population	An organism that must obtain organic nutrients from other organisms
Community	An organism that produces its own organic nutrients from inorganic compounds
Ecosystem	An organism that obtains organic nutrients from decomposing organisms by internal digestion

4.1–4.2 Mode of Nutrition and Energy Flow

Command terms: state and identify

Use the following information about species found in an Australian grassland ecosystem to answer the questions below.

Plants:

Golden Wattle tree (*Acacia pycnantha*)
Lemon-scented Eucalyptus tree (*Eucalyptus citriodora*)
Kangaroo grass (*Themeda australis*) and
Golden Beard grass (*Chrysopogon fallax*)

Animals:

Blue-faced honeyeaters (*Entomyzon cyanotis*) feed on nectar and sap from *E. citriodora*
Termites (*Mastotermes darwiniensis*) feed on *E. citriodora, A. pycnantha* and *T. australis*
Field crickets (*Lepidogryllus comparatus*) feed on *T. australis*
Hairy-nosed wombats (*Lasiorhinus krefftii*) feed on *C. fallax*
Eastern Grey kangaroos (*Macropus giganteus*) feed on *T. australis*
Emus (*Dromaius novaehollandiae*) feed on *A. pycnantha* and *L. comparatus*
Short-beaked echidnas (*Tachyglossus aculeatus*) feed on *M. darwiniensis*

Magpies (*Cracticus tibicen*) feed on *M. darwiniensis* and *L. comparatus*

Frilled-neck lizards (*Chlamydosaurus kingii*) feed on *M. darwiniensis* and *L. comparatus*

Dingoes (*Canis lupus dingo*) feed on *M. giganteus, D. novaehollandiae, L. krefftii* and *C. kingii*

Wedge-tailed eagles (*Aquila audax*) feed on *M. giganteus* and *C. kingii*

Laughing kookaburras (*Dacelo novaeguineae*) feed on *C. kingii*

1. Identify four different food chains from this food web.

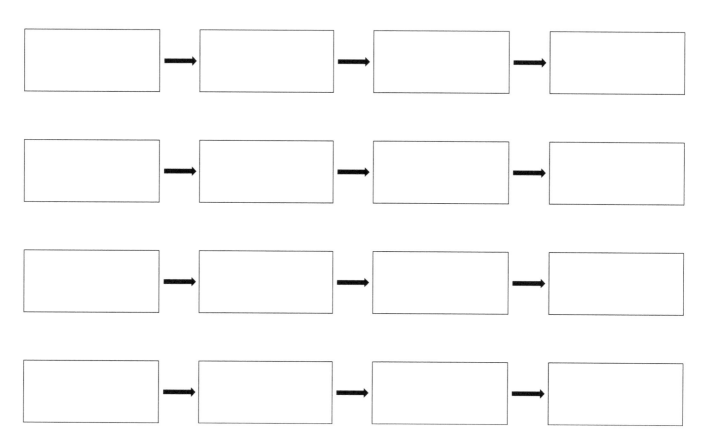

2. Identify the trophic level of each species by completing the table below.

Species	Trophic level
M. darwiniensis	
C. tibicen	
A. pycnantha	
L. krefftii	
C. dingo	
C. kingii	
D. novaehollandiae	

Species	Trophic level
E. cyanotis	
T. australis	
A. audax	
E. citriodora	
M. giganteus	
D. novaeguineae	
C. fallax	
L. comparatus	
T. aculeatus	

3. Draw an energy pyramid using one food chain as an example.

4.1 Chi-Squared Test

Command terms: measure, calculate, estimate, determine and predict

The chi-squared test is used to test if an observed frequency is statistically and significantly different to the expected frequency. In biology, a difference of 0.05 is said to be statistically significant because it gives us 95 per cent confidence in that difference being significant. The number of degrees of freedom is equal to the number of possible outcomes minus 1. There are two possible hypotheses; two species are distributed independently (H_0) or two species are either positively or negatively associated (H_1).

The formula for the chi-squared test is:

$$X^2 = \Sigma \frac{(O - E)^2}{E}$$

where

X^2 = the test statistic

O = the observed values

E = the expected values

Σ = sum of

Data were collected in a section of the Daintree Rainforest in tropical far north Queensland, Australia. Two plants in the area were surveyed using 50 quadrats and their numbers are shown in the contingency table below. *Asplenium australasicum* (Birds Nest Fern) is found growing high up on trees such as *Athertonia diversifolia* (Atherton Oak) as well as low lying on rocks.

	Asplenium australasicum present	*Asplenium australasicum* absent	Row totals
Athertonia diversifolia present	24	4	28
Athertonia diversifolia absent	13	9	22
Column totals	37	13	50

To use the chi-squared test:

1. Calculate the expected values, assuming there is **no** association between the two species.

_____

_____

2. Calculate the number of degrees of freedom.

_____

_____

3. Find the critical value for chi-squared at a significance level of 0.05 (5 per cent) using the chi-squared distribution table below.

_____

_____

Percentage Points of the Chi-Square Distribution

Degrees of freedom	Probability of a larger value x^2								
	0.99	0.95	0.90	0.75	0.50	0.25	0.10	0.05	0.01
1	0.000	0.004	0.016	0.102	0.455	1.32	2.71	3.84	6.63
2	0.020	0.103	0.211	0.575	1.386	2.77	4.61	5.99	9.21
3	0.115	0.352	0.584	1.212	2.366	4.11	6.25	7.81	11.34
4	0.297	0.711	1.064	1.923	3.357	5.39	7.78	9.49	13.28
5	0.554	1.145	1.610	2.675	4.351	6.63	9.24	11.07	15.09
6	0.872	1.635	2.204	3.455	5.348	7.84	10.64	12.59	16.81
7	1.239	2.167	2.833	4.255	6.346	9.04	12.02	14.07	18.48
8	1.647	2.733	3.490	5.071	7.344	10.22	13.36	15.51	20.09
9	2.088	3.325	4.168	5.899	8.343	11.39	14.68	16.92	21.67
10	2.558	3.940	4.865	6.737	9.342	12.55	15.99	18.31	23.21
11	3.053	4.575	5.578	7.584	10.341	13.70	17.28	19.68	24.72
12	3.571	5.226	6.304	8.438	11.340	14.85	18.55	21.03	26.22
13	4.107	5.892	7.042	9.299	12.340	15.98	19.81	22.36	27.69
14	4.660	6.571	7.790	10.165	13.339	17.12	21.06	23.68	29.14
15	5.229	7.261	8.547	11.037	14.339	18.25	22.31	25.00	30.58
16	5.812	7.962	9.312	11.912	15.338	19.37	23.54	26.30	32.00
17	6.408	8.672	10.085	12.792	16.338	20.49	24.77	27.59	33.41
18	7.015	9.390	10.865	13.675	17.338	21.60	25.99	28.87	34.80
19	7.633	10.117	11.651	14.562	18.338	22.72	27.20	30.14	36.19
20	8.260	10.851	12.443	15.452	19.337	23.83	28.41	31.41	37.57
22	9.542	12.338	14.041	17.240	21.337	26.04	30.81	33.92	40.29
24	10.856	13.848	15.659	19.037	23.337	28.24	33.20	36.42	42.98
26	12.198	15.379	17.292	20.843	25.336	30.43	35.56	38.89	45.64
28	13.565	16.928	18.939	22.657	27.336	32.62	37.92	41.34	48.28
30	14.953	18.493	20.599	24.478	29.336	34.80	40.26	43.77	50.89
40	22.164	26.509	29.051	33.660	39.335	45.62	51.80	55.76	63.69
50	27.707	34.764	37.689	42.942	49.335	56.33	63.17	67.50	76.15
60	37.485	43.188	46.459	52.294	59.335	66.98	74.40	79.08	88.38

Source: Plant and Soil Sciences eLibrary, University of Nebraska-Lincoln, USA, 2015, http://passel.unl.edu/pages/informationmodule.php?idinformationmodule=1130447119, used with permission.

4. Calculate chi-squared using the formula.

_____

_____

_____

5. State the two hypotheses (H_0 and H_1) for this data and evaluate them using the value you have calculated for chi-squared.

_____

_____

_____

6. Explain methods that should be used to collect the data in a study like this.

_____

_____

_____

7. Why is it important to base sampling on random numbers.

_____

_____

_____

8. Suggest reasons for any association between *A. australasicum* and *A. diversifolia.*

_____

_____

_____

9. What are some of the factors that could have affected the distribution of the two species in the Daintree Rainforest?

_____

_____

_____

4.2 Energy Flow through Ecosystems

Command terms: describe, outline and explain

Number the structures below in the correct order to show the flow of energy through ecosystems.

Primary consumers obtain energy from plant food   _____

Tertiary consumers obtain energy by eating other animals   _____

The sun provides light energy   _____

Decomposers obtain energy from dead bodies and waste   _____

Secondary consumers obtain energy by eating other animals   _____

Heat energy leaves the ecosystem   _____

Quaternary consumers obtain energy by eating other animals   _____

Primary producers convert sunlight into chemical energy by photosynthesis   _____

4.3 The Carbon Cycle

Command terms: draw, label, annotate and construct

Label the arrows to represent the flow of carbon through a terrestrial ecosystem.

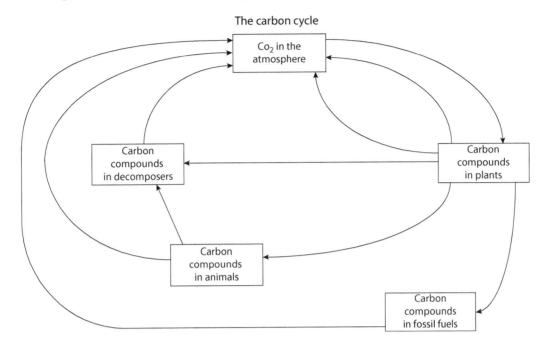

The carbon cycle

4.3 The Carbon Cycle

Command terms: state, list and identify

Place the following processes involved in the carbon cycle in the correct order from the least carbon fluxes to the most carbon fluxes.

Ocean loss	_____
Burial in marine sediments	_____
Photosynthesis	_____
Ocean uptake	_____
Fossil fuel combustion	_____
Cellular respiration	_____
Deforestation	_____

4.3 Forms of Carbon

Command terms: state and identify

The following table summarises the forms of carbon that occur throughout the carbon cycle.

Form	Formula	Found
Calcium carbonate	$CaCO_3$	Rocks, limestone, soil, shells and coral
Carbon dioxide	CO_2	Atmosphere and dissolved in water
Hydrogen carbonate ion	HCO_3^-	Dissolved in water
Carbonic acid	H_2CO_3	Dissolved in water
Organic carbon	C	Tissues of living organisms
Methane	CH_4	Decaying organic matter with no oxygen present

4.3 Methane Production

Command terms: list, state and identify

Determine which of the following conditions are necessary for the production of methane by annotating each statement with 'true' or 'false'

	True/False
Methane is produced from organic matter	_____
Bacteria are involved in the production of methane	_____
Oxygen is a requirement of methane production	_____
Methane accumulates in the ground	_____
Accumulation of methane causes peat formation	_____
Carbon dioxide is a requirement of methane production	_____
Anaerobic conditions are needed for methane production	_____
Oxygen is a by-product of methane production	_____
Methane present in the atmosphere is converted to water	_____
Methane is a by-product of anaerobic respiration	_____
Decay of inorganic matter produces methane	_____
Methane is produced from carbon dioxide, hydrogen and acetate	_____

4.4 Greenhouse Gases

Command terms: list, state and outline

Complete the table below to summarise the impacts of the greenhouse gases listed. Rank them in the order of significance.

Rank	Greenhouse gas	Impact
	Methane	
	Water vapour	
	Carbon dioxide	
	Nitrous oxide	

4.4 The Greenhouse Effect

Command terms: list, state and identify

Determine if the following statements about the greenhouse effect are 'true' or 'false'

	True/False
Greenhouse gases trap heat near the surface of the Earth	_____
The concentration of a greenhouse gas determines it's warming impact on the Earth	_____
Greenhouse gases reabsorb longer wave radiation	_____
Short-wave radiation is absorbed from the sun	_____
Water vapour is one of the most significant greenhouse gases	_____
The Earth would be better off without any greenhouse gases	_____
The increase in CO_2 emissions is a major factor contributing to the enhanced greenhouse effect	_____
The greenhouse effect is caused when longer wave radiation is prevented from escaping the atmosphere	_____
Greenhouse gases raise the temperature of the Earth around 30°C	_____
Oxygen and nitrogen are both greenhouse gases	_____
The enhanced greenhouse effect causes global warming	_____
Climate change is caused only by the enhanced greenhouse effect	_____
Together, all greenhouse gases make up <1% of the Earth's atmosphere	_____
Both man-made and natural processes contribute to the enhanced greenhouse effect	_____

EVOLUTION AND BIODIVERSITY (TOPIC 5)

5.1 Evidence for Evolution

Command terms: define and state

Link the following terms with their meanings.

Adaptive radiation	A structure similar in form and/or function found in different animals suggesting descent from a common ancestor.
Continuous variation	The evolution of a number of divergent species, occupying a different environment, from a common ancestor.
Gradual divergence	The breeding of plants or animals by humans to produce desirable traits.
Homologous structures	The gradual accumulation of variation between populations and leading to evolution.
Selective breeding	Variation of a trait that has no limit on the value that can occur within a population. A complete range of measurements can occur from one extreme to the other.

5.1 Limb Comparison

Command terms: label, state, annotate, identify and compare

The structure of the pentadactyl limb can be compared in animals with different methods of locomotion. The following diagram depicts the structure of this limb in a number of mammals.

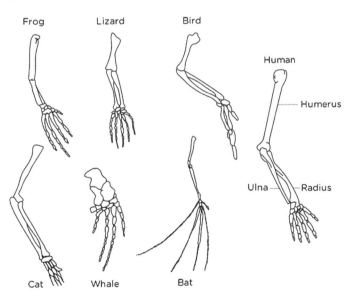

1. Colour code each of the different bones found in the limb.
   - ○ Humerus
   - ○ Ulna
   - ○ Radius
   - ○ Phalanges

2. Annotate the diagram to state how each limb is used.

5.2 Natural Selection

Command terms: state, describe, outline and explain

Fill in the blanks to complete the sentences below that outline Charles Darwin's theory of Natural Selection.

Individuals within a population show natural _____ in traits.

This comes about through _____, _____ and sexual reproduction.

The _____ selects those traits that are best suited.

Species tend to produce _____ than the environment can support.

This brings about _____ between individuals.

The individuals with these traits are able to _____ and therefore _____.

This allows the favourable traits to be passed on to the _____, which therefore have an increased chance of _____

Over time, the _____ in the population changes and eventually _____ _____ occurs.

5.2 Galápagos Finches

Command terms: state, describe, identify, compare, discuss and suggest

1. Geospiza magirostris   2. Geospiza fortis
3. Geospiza parvula   4. Certhidea olivacea

Finches from Galapagos Archipelago

Source: John Gould, 14 September 1804–3 February 1881 *Galapagos finches* (Public domain), via Wikimedia Commons, https://commons. wikimedia.org/wiki/Category:Darwin%27s_finches#/media/File:Charles_Darwin,_Journal_of_Researches..._Wellcome_L0026712.jpg.

The image above shows 4 of the 14 different species of finches found on the islands in the Galápagos archipelago. The finches each have different size and shape beaks.

1. Suggest the diet of each of the four species of finch pictured.

   _____

   _____

   _____

2. Discuss how the different conditions on each of the islands of the Galápagos could have contributed to the difference in beak size and structure.

   _____

   _____

   _____

3. Explain how Darwin's theory of Natural Selection can be applied to the development of fourteen different species of finches on the Galápagos Islands.

   _____

   _____

   _____

   _____

5.2 Antibiotic Resistance of Bacteria

Command terms: state, describe, outline and explain

Fill in the blanks to complete the sentences below that outline the resistance to multiple antibiotics by bacteria.

Antibiotics impose a _____.

Normally, bacteria are killed by _____ as their cell walls are ruptured.

Bacteria have natural variation with some individuals having _____ _____ to antibiotics.

This resistance is caused by a _____ mutation.

The bacteria with resistance produce an _____ that works against the antibiotic.

If this resistance is _____, it is passed on to offspring.

Offspring reproduce and ultimately form bacterial colonies that are _____ to the antibiotic.

This renders the antibiotic _____.

Humans have added to the problem of antibiotic resistance by doctors _____, doctors prescribing for _____ infections, patients not completing _____ and adding antibiotics to _____.

5.3 Features of Plant Phyla

Command terms: list, state and identify

The table below summarises key features for plant phyla.

Phylum	Example	Features
Bryophyta	Mosses	Small, reproduce by spores, non-woody stems, rhizoids instead of true roots, no leaf cuticles, no xylem or phloem, no cambium, no seeds, no fruit
Filicinophyta	Ferns	True roots, stems and leaves, new leaves unroll, underground creeping stem, xylem and phloem present, no cambium, no seeds, no fruit
Coniferophyta	Conifers	True roots, stems and leaves, woody stems, produce seeds carried in cones, leaves are long and thin with cuticles, have xylem and phloem, have cambium, no fruit
Angiospermophyta	Flowering plants	True roots, stems and leaves, have flowers, seeds are ovaries that become fruit, leaf blades and stalks with veins visible on lower surface, have xylem and phloem, have cambium

5.3 Features of Animal Phyla

Command terms: list, state and identify

The table below summarises key features for animal phyla.

Phylum	Example	Features
Porifera	Sponges	Simple body, do not move, no mouth but instead have holes for water to be pumped into body, filter water to obtain food
Cnidaria	Jellyfish	Tentacles with cnidocytes, radial symmetry, no shell, only one opening to cavity (no mouth/anus)
Platyhelmintha	Flat worms	Soft-flattened body, bilateral symmetry, hollow space in the centre of body with only one opening (no mouth/anus)
Annelida	Round worms	Segmented bodies with bristles for movement, have mouth and anus, no legs or shell
Mollusca	Snails, squid, clams, slugs	Bilateral symmetry, no tentacles, soft unsegmented bodies, have mouth and anus, may have a hard shell
Arthropoda	Animals with jointed legs	Exoskeleton made of chitin, segmented body, appendages on each segment, at least three pairs of jointed legs
Chordata	Animals with a backbone	Single hollow nerve cord (brain and spinal cord in vertebrates), notochord, pharyngeal slits, postanal tail/tailbone, segmentation

5.3 Features of Animal Classes

Command terms: list, state and identify

The table below summarises key features for animal classes.

Class	Example	Features
Birds	Sulphur-crested cockatoo (*Cacatua galerita*)	Warm-blooded (endothermic), light skeleton with air sacs, feathers for body insulation and flight, wings, bill/beak, bipedal, egg-layers
Mammals	Red kangaroo (*Macropus rufus*)	Warm-blooded (endothermic), mammary glands for nursing young, body covered in hair/fur, lower jaw made of a single bone, diphyodonty (tooth replacement), three middle ear bones, diaphragm, four-chambered heart

Class	Example	Features
Amphibians	Green tree frog (*Litoria caerulea*)	Cold-blooded (exothermic), live in water and on land, body covered in permeable skin with no hair or scales, gills, go through metamorphis, three-chambered heart, egg-layers
Reptiles	Salt water crocodile (*Crocodylus porosus*)	Almost all are cold-blooded (exothermic), body covered in scales, have lungs for breathing, most lay eggs, four-chambered heart
Fish	Clown Anemonefish (*Amphiprion percula*)	Almost all are cold-blooded (exothermic), body covered in scales, fins for movement, breathe using gills, live in water, egg-layers

5.3 Classification of Plants and Animals

Command terms: list, state and identify

The table below gives the full classification for one plant species and one animal species.

	Plant example	Animal example
	Sturt's Desert Pea (*Swainsona formosa*)	Hairy Nosed Wombat (*Lasiorhinus krefftii*)
Domain	Eukaryota	Eukaryota
Kingdom	Plantae	Animalia
Phylum	Angiospermophyta	Chordata
Class	Magnoliopsida	Mammalia
Order	Fabales	Diprotodontia
Family	Fabaceae	Vombatidae
Genus	Swainsona	Lasiorhinus
Species	Formosa	Krefftii

5.3 Dichotomous Keys

Command terms: distinguish, identify, construct and determine

Construct a dichotomous key to classify the individual beetles shown below.

Source: Georgiy Jacobson, 1915, *Beetles Russia and Western Europe* (Public domain) via Wikimedia Commons, http://commons.wikimedia.org/
wiki/Category:Georgiy_Jacobson._Beetles_Russia_and_Western_Europe_(extracted_images).

5.4 Human and Primate Cladogram

Command terms: state, identify, analyse, deduce and determine

The following is a cladogram showing the evolutionary relationship between members of the order Primates. Primates are members of the phylum Chordata and class Mammalia and include humans.

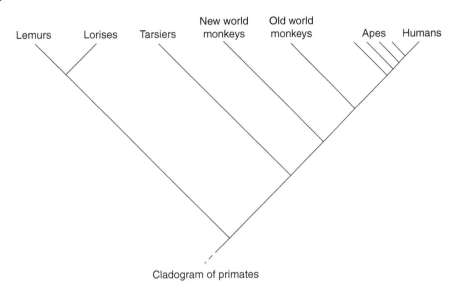

Cladogram of primates

Source: Petter Bøckman, 2011, *Primate Cladogram*, http://commons.wikimedia.org/wiki/File:Primate_cladogram.jpg.

Use the cladogram to answer the following questions.

1. Deduce, using evidence from the cladogram, whether humans are more closely related to lemurs or lorises.

   _____

   _____

2. Identify whether lemurs and tarsiers or new world monkeys and old world monk.1eys are more closely related.

   _____

   _____

3. Are humans and lorises considered to be part of the same clade? Justify your answer.

   _____

   _____

   _____

.

6.1 The Human Digestive System

Command terms: draw, label, annotate and identify

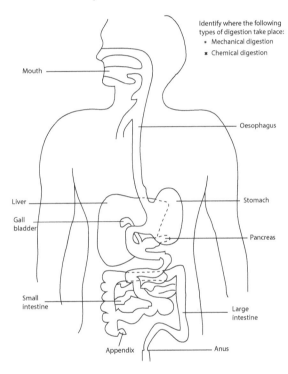

Identify where the following types of digestion take place:
- • Mechanical digestion
- ✖ Chemical digestion

Mouth

Oesophagus

Liver

Stomach

Gall bladder

Pancreas

Small intestine

Large intestine

Appendix

Anus

6.1 Structure and Function of the Small Intestine

Command terms: state and outline

Complete the table below to outline the function of the named structures of the small intestine.

Structure	Function
Circular muscle	
Longitudinal muscle	
Mucosa	
Epithelium	
Enzymes	
Villi	
Length	

6.1 Nutrient Transport

Command terms: state and identify

Complete the table below to show how different nutrients are transported across the membrane of the small intestine either by **(a) simple diffusion, (b) facilitated diffusion, (c) active transport or (d) exocytosis.**

Nutrient	Mode of membrane transport
Amino acids	
Fatty acids Monoglycerides Glycerol	
Glucose Fructose Galactose Other monosaccharides	
Nitrogen bases from nucleotides	
Mineral ions such as calcium, potassium and sodium	
Vitamins such as ascorbic acid	
Water	

6.2 The Human Heart

Command terms: draw, label, annotate and identify

Use coloured pencils to colour the parts of the heart where oxygenated (red) and deoxygenated (blue) blood is found.

Use arrows to show the direction of blood flow through the heart.

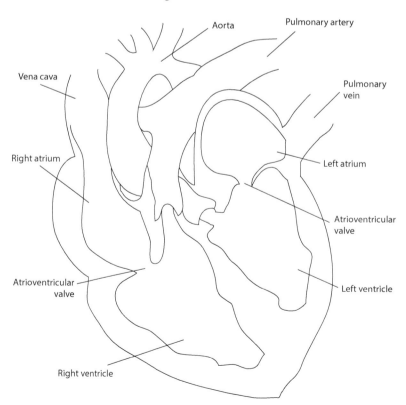

6.2 Blood Vessels

Command terms: distinguish, compare, compare and contrast

Complete the table below to summarise the structure of human blood vessels.

	Veins	Arteries	Capillaries
Blood pressure		Convey blood at high pressure	
Direction of blood flow	Carry blood towards the heart[a]		
Wall thickness			Walls have only one layer of cells to allow rapid diffusion
Lumen		Narrow lumen to maintain the high pressure	
Valves	Valves prevent backflow of blood		
Structure of walls		Muscle cells and elastic fibres in walls	Permeable walls allow exchange of materials
Secondary structure	Venule	Arteriole	n/a

n/a, not applicable.

[a] Except the pulmonary vein which carries blood away from the heart.

6.2 Blood Flow through the Heart

Command terms: describe, identify and outline

Number the structures below in the correct order to show the path of blood flow through the heart and around the body. The right atrium has been completed for you.

Right atrium	1
Left atrium	
Right ventricle	
Left ventricle	
Lungs	
Aorta	
Vena cava	
Pulmonary artery	
Pulmonary vein	
Atrio-ventricular (AV) valves	
Semilunar valves	

6.2 Pressure Changes in the Heart

Command terms: state, annotate, identify and outline

The table below shows the timeline and events occurring during one cardiac cycle.

Time (secs)	0.00-0.10	0.10 - 0.15	0.15-0.40	0.40 - 0.45	0.45-0.80
atrium	contracts	relaxed			
AV valve	open	closed		open	
ventricle	relaxed	contracts		relaxed	
SL valve	closed	open		closed	
artery	diastolic	systolic		diastolic	

1. Annotate the table using arrows to show the direction of blood flow and when this blood flow occurs.
2. Outline reasons for the change in pressure at the following points in the cardiac cycle.

   a. Atrial contraction

   _____

   _____

   b. Closing of semilunar valves

   _____

   _____

   c. Closing of AV valves

   _____

   _____

   d. Ventricular contraction

   _____

   _____

   e. Opening of semilunar valves

   _____

   _____

   f. Opening of AV valves

   _____

   _____

6.2–6.3 Atherosclerosis and Coronary Thrombosis

Command terms: list, state and compare

The table below summarises the causes or risk factors and consequences of the occlusion of coronary arteries known as atherosclerosis and the resultant blood clot formation known as coronary thrombosis.

Causes/risk factors	Consequences
Chronic high blood pressure (often due to smoking or stress)	
High blood concentration of LDL	Pain (angina)
High blood cholesterol concentration	Increased heart rate
	Hardened arteries leading to blockages
Chronic high blood sugar levels (often due to diabetes or obesity)	Blood clot formation
	Myocardial infarction (heart attack)
High consumption of trans fats	
Lack of exercise	

LDL, low-density lipoprotein.

6.3 Antibiotics

Command terms: list, state and identify

Determine which of the following statements about antibiotics are true or false.

	True/False
Antibiotics are effective in blocking processes that occur in both eukaryotic and prokaryotic cells	_____
Antibiotics function by lysing the cell membranes of their target cells	_____
Antibiotics are only useful for bacterial infections	_____
Antibiotics are useful for both viral and bacterial infections	_____
Antibiotics are useful for bacterial cells as they are identified by the body as non-self	_____
Antibiotics inhibit the formation of the cell wall and protein synthesis	_____
The source of many antibiotics is fungi	_____
Antibiotics cannot be used to treat bacterial infections	_____
Antibiotics do not affect viruses as they lack a metabolism	_____
Antibiotics can cause harm by killing off helpful bacteria	_____

6.3 Human Immunodeficiency Virus

Command terms: list, state and outline

The table below summarises the transmission and effects of Human Immunodeficiency Virus (HIV).

Transmission	Effects
Sexual intercourse (man to woman or man to man)	Loss or ineffectiveness of antibody production
Across the placenta from mother to foetus	Weakened or disabled immune system, more susceptible to opportunistic pathogens
Via breast milk from mother to foetus	
Blood transfusions involving infected blood or blood products	Secondary infections such as pneumonia have a more damaging effect
From needle-stick injuries or shared needles by intravenous drug users	Acquired immune deficiency syndrome (AIDS)

6.3 Blood Clotting

Command term: outline

The flowchart shows the process of blood clotting after a wound or injury to a blood vessel.

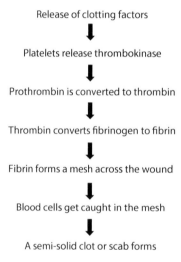

Release of clotting factors

⬇

Platelets release thrombokinase

⬇

Prothrombin is converted to thrombin

⬇

Thrombin converts fibrinogen to fibrin

⬇

Fibrin forms a mesh across the wound

⬇

Blood cells get caught in the mesh

⬇

A semi-solid clot or scab forms

6.4 Mechanism of Ventilation

Command terms: identify, distinguish and compare

Complete the following table to summarise the mechanism of ventilation in humans.

	Inspiration	Expiration
Diaphragm		
Abdominal muscles		
External intercostal muscles		
Internal intercostal muscles		
Rib cage		
Volume of thoracic cavity		
Air pressure in lungs		
Air movement		

6.4 Lung Cancer and Emphysema

Command terms: list, state and compare

The table below summarises the causes and consequences of two lung diseases, lung cancer and emphysema.

Lung cancer		Emphysema	
Causes	**Consequences**	**Causes**	**Consequences**
Tobacco smoking		Tobacco smoking	
Air pollution, especially diesel exhaust	Difficulty breathing, shortness of breath, coughing and coughing up blood, chest pain, fluid in chest, appetite loss, general fatigue, metastasis	Marijuana smoking	Chronic shortness of breath, rapid breathing, coughing and wheezing, lack of energy, tiredness
Passive smoking		Alpha 1-antitrypsin (A1AT) deficiency (genetic)	
Radon gas		Passive smoking	
Asbestos, silica		Air pollution	

6.5 Synaptic Transmission

Command terms: label, state, annotate and describe

Annotate the following diagram to describe the steps occurring during the synaptic transmission of an action potential.

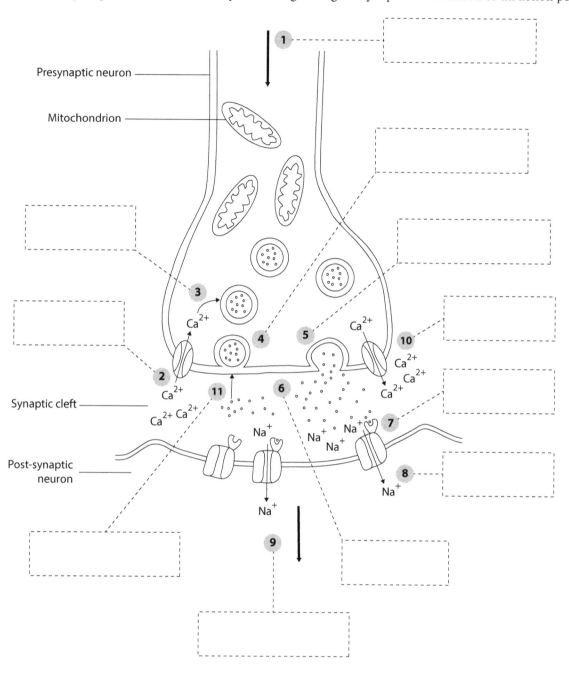

6.5 Synaptic Transmission

Command terms: describe, outline and explain

Number the steps below in the correct order to show the process of synaptic transmission.

Sodium ions flow into the post-synaptic neuron   _____

Calcium ions enter the presynaptic neuron   _____

Neurotransmitter diffuses across the synaptic cleft   _____

Neurotransmitter binds to receptors in the membrane of the post-synaptic neuron   _____

Neurotransmitter travels through ion channels in the post-synaptic membrane   _____

Neurotransmitter is released from synaptic vesicles by exocytosis   _____

The membrane of the post-synaptic neuron is depolarised   _____

An action potential is initiated   _____

An action potential arrives at the presynaptic knob   _____

Neurotransmitter returns to the presynaptic neuron   _____

Ion channels open on the post-synaptic neuron   _____

The nerve impulse travels down the presynaptic neuron to the presynaptic knob   _____

Synaptic vesicles containing neurotransmitter fuse to the presynaptic membrane   _____

Neurotransmitter is removed from the synaptic cleft   _____

Calcium ions are pumped back into the synaptic cleft   _____

6.6 Type I and Type II Diabetes

Command terms: outline, distinguish, compare, compare and contrast

Complete the following table to compare the causes and treatment of Type I and Type II diabetes.

	Type I diabetes	Type II diabetes
Onset		
Production of insulin		
Sensitivity to insulin		
Autoimmune involvement		
Genetic predisposition		
Diet and lifestyle		
Control of disease		

6.6 Human Reproductive Systems

Command terms: label, annotate, identify and outline

The diagrams show the structure of female and male reproductive systems. Label the parts indicated on both diagrams and add annotations to outline function of named structures.

Female

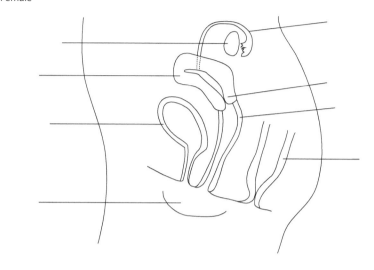

Male

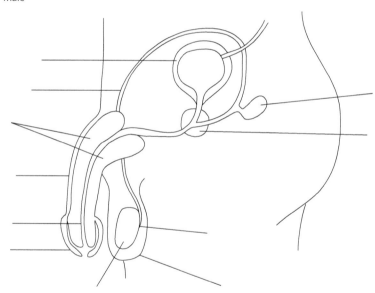

6.6 Positive and Negative Feedback

Command terms: distinguish, compare, compare and contrast

Below are summaries of positive and negative feedback mechanisms.

Negative feedback

- Most homeostatic control mechanisms occur as a result of negative feedback.
- In such mechanisms, the output of the system shuts off the original stimulus or reduces its intensity.

- These mechanisms cause the variable to change in a direction opposite to that of the initial change, returning it back to its 'ideal' value.
- All negative feedback mechanisms have the same goal: to prevent sudden severe changes within the body.
- Examples include:

  - Regulation of body temperature
  - Withdrawal reflex (from a painful stimulus)
  - Control of blood glucose levels by pancreatic hormones
  - The effect of progesterone on the pituitary gland to block the release of follicle-stimulating hormone (FSH)
  - The involvement of anti-diuretic hormone (ADH) in osmoregulation
  - Regulation of blood pH levels during exercise

Positive feedback

- In positive feedback mechanisms, the result or response enhances or exaggerates the initial stimulus so that the activity or output is accelerated.
- This feedback mechanism is said to be 'positive' as the change that occurs proceeds in the same direction as the initial disturbance, causing the variable to deviate further from its original value or range.
- Positive feedback mechanisms usually control infrequent events that do not require constant adjustment.
- Examples include:

  - Blood clotting
  - Enhancement of labour contractions during birth (oxytocin)
  - The effect of oestrogen on luteinizing hormone (LH) to trigger ovulation.

6.6 Positive and Negative Feedback

Command terms: state and identify

Organise the following examples under the headings of either positive feedback or negative feedback.

Examples

Regulation of body temperature
Milk production for lactating mothers
Gene activation
Control of blood glucose levels by pancreatic hormones
End-product inhibition in enzymes
Generation of nerve signals (action potentials)
Regulation of oxygen levels in the blood
Withdrawal reflex from a painful stimulus
Regulation of blood pH levels during exercise
Effect of progesterone on the pituitary gland to block the release of FSH
Effect of oestrogen on LH to trigger ovulation
Enhancement of labour contractions by oxytocin during birth
Control of blood pressure (heart rate plus vasoconstriction/vasodilation)
Blood clotting
Involvement of ADH in osmoregulation
The sensation of thirst
Digestion of proteins in the stomach triggering the secretion of hydrochloric acid and pepsin

Positive feedback	Negative feedback

6.6 Hormone Roles

Command terms: list, state and identify

The following table summarises key human hormones and their role in the body.

Hormone	Secreted by	Main target organ	Type	Role
Insulin	Pancreatic β cells	Liver (most tissues)	Peptide	Controls blood glucose concentration
Glucagon	Pancreatic α cells	Liver	Peptide	Controls blood glucose concentration
Thyroxin	Thyroid gland	Skeletal and muscles (most cells)	Amino acid	Regulates metabolic rate and help control body temperature
Leptin	Adipose tissue	Hypothalamus	Peptide	Inhibits appetite
Melatonin	Pineal gland	Brain	Amino acid	Controls circadian rhythms
FSH	Anterior pituitary	Ovaries, testes	Peptide	Stimulates follicle development in the ovary Stimulates the secretion of oestrogen by the ovary Stimulates meiosis in developing spermatocytes
LH	Anterior pituitary	Ovaries, testes	Peptide	Causes ovulation Causes development of the corpus luteum Causes the secretion of progesterone Stimulates Leydig cells to produce testosterone
Oestrogen	Ovaries, testes, placenta	Ovaries	Steroid	Prenatal development of female reproductive organs Development of female secondary sex characteristics Stimulates repair and thickening of the uterine lining Stimulates secretion of LH and FSH
Progesterone	Ovaries, adrenal gland, placenta	Uterus, mammary glands	Steroid	Prenatal development of female reproductive organs Development of female secondary sex characteristics Causes thickening of the uterine lining Inhibits secretion of LH and FSH
Testosterone	Testes (Leydig cells)	Testes	Steroid	Prenatal development of male genitalia Development of male secondary sex characteristics Stimulation of sperm production Maintenance of sex drive
Epinephrine	Adrenal glands	Heart, lungs	Catecholamine	Increases heart rate and respiratory rate to prepare for vigorous physical activity

Hormone	Secreted by	Main target organ	Type	Role
Additional Higher Level (AHL)				
ADH	Posterior pituitary	Kidney (collecting ducts)	Peptide	Regulates water levels in the body
Gastrin	Stomach, duodenum, pancreas	Stomach (parietal cells)	Peptide	Stimulates the secretion of HCL in the stomach
Oxytocin	Posterior pituitary	Uterus mammary glands	Peptide	Stimulates uterine contractions during labour Control of milk secretion
hCG	Placenta	Placenta	Peptide	Matures the follicles in the ovary Promotes maintenance of corpus luteum during pregnancy
Prolactin	Anterior pituitary, uterus	Mammary glands, ovaries	Peptide	Control of milk secretion
Growth hormone	Anterior pituitary	Liver, bones, muscles	Peptide	Stimulates cell growth and cell reproduction
Plant hormones				
Auxin	Apical bud or tip	Stem		Influences cell growth rates by changing the pattern of gene expression

NUCLEIC ACIDS (TOPIC 7) HIGHER LEVEL

7.1 DNA Replication

Number the following steps of DNA replication in prokaryotes in the correct order and identify which of the steps occurs on either the sense strand or the antisense strand.

Step	Number	Sense strand or antisense strand?
Okazaki fragments are joined together		
RNA primer is removed		
RNA primer is created		
Hydrogen bonds between base pairs are broken		
2 Phosphates are removed from each dNTP		
The DNA double helix is unwound		
Okazaki fragments are formed		
RNA nucleotides are replaced with DNA nucleotides		
Complementary base pairing occurs		
dNTPs are added to the growing strand		
DNA double helix reforms		
The single stranded binding (SSB) proteins maintain strand separation		
RNA primer binds to the old DNA strand		

7.1 DNA Replication – Concept Map

Command terms: define, list, state and identify

Word bank

double helix
DNA polymerase I
deoxyribose
deoxyribonucleotide
nitrogen base
DNA nucleotides
RNA nucleotides
semi-conservative
DNA polymerase III
Okazaki fragments
DNA replication
hydroxyl group
discontinuous
lagging strand
adenine

eukaryotes
RNA primer
pyrimidine
phosphate
helicase
leading strand
nucleus
proteins
double helix
continuous
enzymes
nucleosomes
DNA ligase
RNA primase

guanine
DNA
one
purine
sugar
5'
energy
two
thymine
histones
3'
dNTTs
H-bonds
cytosine

7.1 DNA Replication

Command terms: state, outline, describe and explain

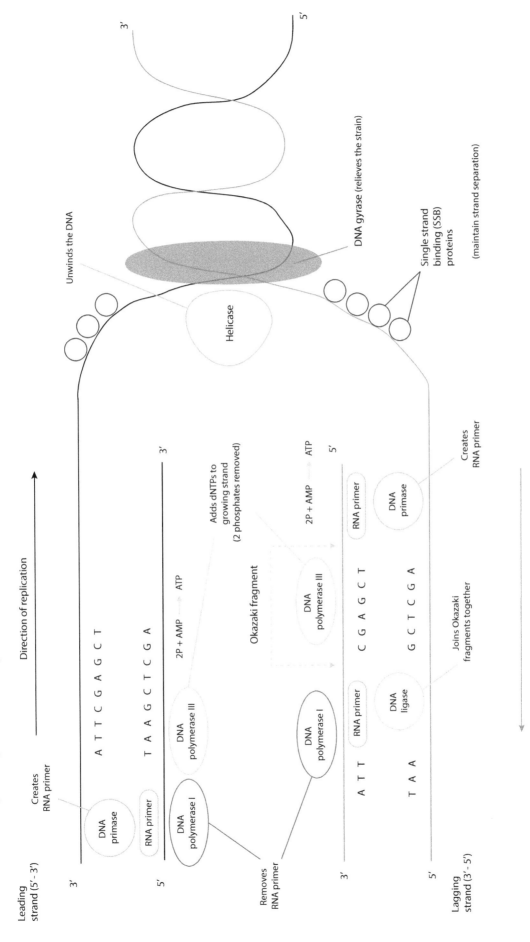

Leading strand (5' - 3')

Direction of replication

Creates RNA primer

DNA primase

3'

RNA primer

A T T C G A G C T

5'

T A A G C T C G A

DNA polymerase I

DNA polymerase III

2P + AMP → ATP

Removes RNA primer

Adds dNTPs to growing strand (2 phosphates removed)

3'

Okazaki fragment

DNA polymerase III

2P + AMP → ATP

DNA polymerase I

RNA primer

DNA ligase

Joins Okazaki fragments together

A T T

T A A

C G A G C T

G C T C G A

RNA primer

DNA primase

Creates RNA primer

3'

5'

Direction of replication

Lagging strand (3' - 5')

Unwinds the DNA

Helicase

DNA gyrase (relieves the strain)

Single strand binding (SSB) proteins

(maintain strand separation)

3'

5'

3'

5'

7.1–7.2 Nucleosomes

Command terms: label, annotate, identify and outline

The diagram shows the structure of a nucleosome.

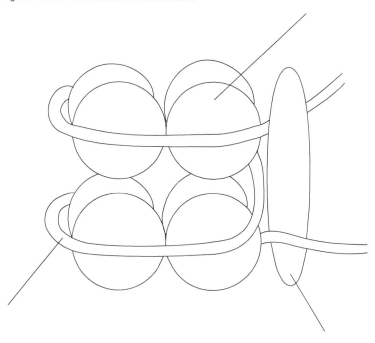

1. Label the indicated parts.

2. Outline the function of nucleosomes in the packaging of eukaryotic DNA.

_____
_____
_____
_____

3. Outline the role of nucleosomes in the regulation of transcription in eukaryotes.

_____
_____
_____
_____

7.2 Transcription

Command terms: state, describe, outline and explain

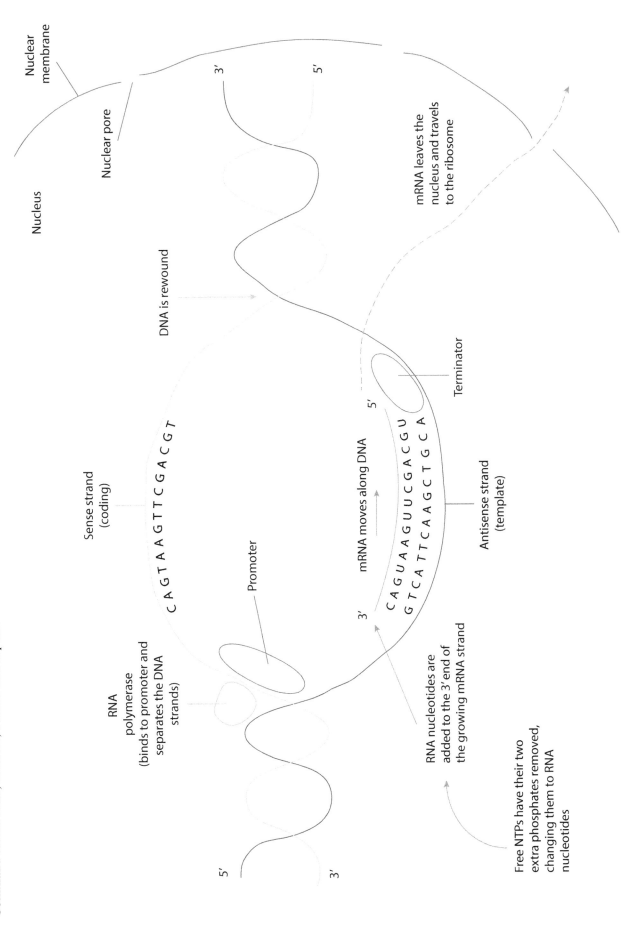

Nuclear membrane

Nucleus

Nuclear pore

3'

5'

mRNA leaves the nucleus and travels to the ribosome

DNA is rewound

Sense strand (coding)

C A G T A A G T T C G A C G T

Promoter

RNA polymerase (binds to promoter and separates the DNA strands)

3'   mRNA moves along DNA

C A G U A A G U U C G A C G U
G T C A T T C A A G C T G C A

5'

Terminator

Antisense strand (template)

RNA nucleotides are added to the 3' end of the growing mRNA strand

5'

3'

Free NTPs have their two extra phosphates removed, changing them to RNA nucleotides

7.3 Translation

Command terms: state, describe, outline and explain

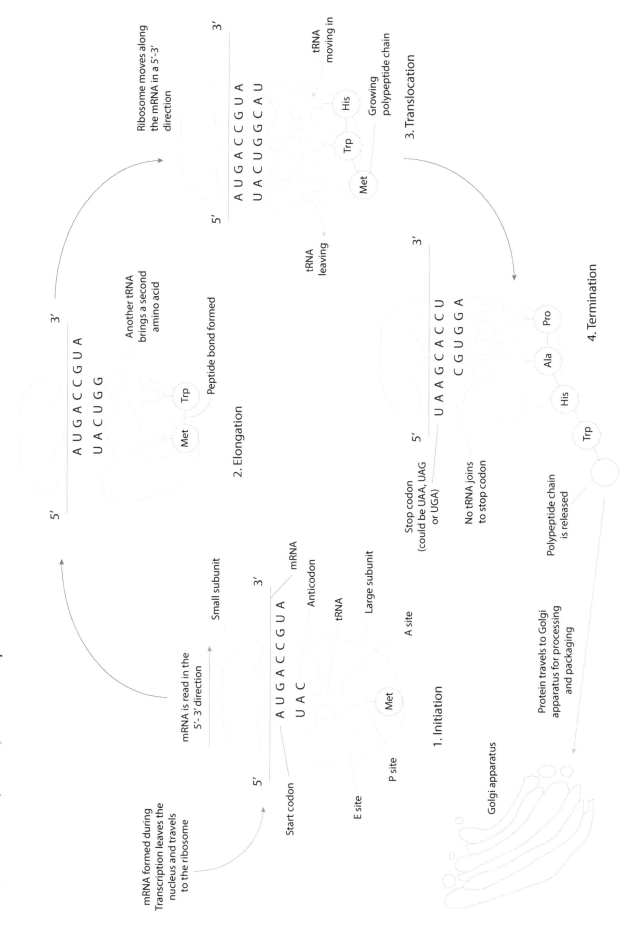

7.2, 7.3 Protein Synthesis – Concept Map

Command terms: define, list, state and identify

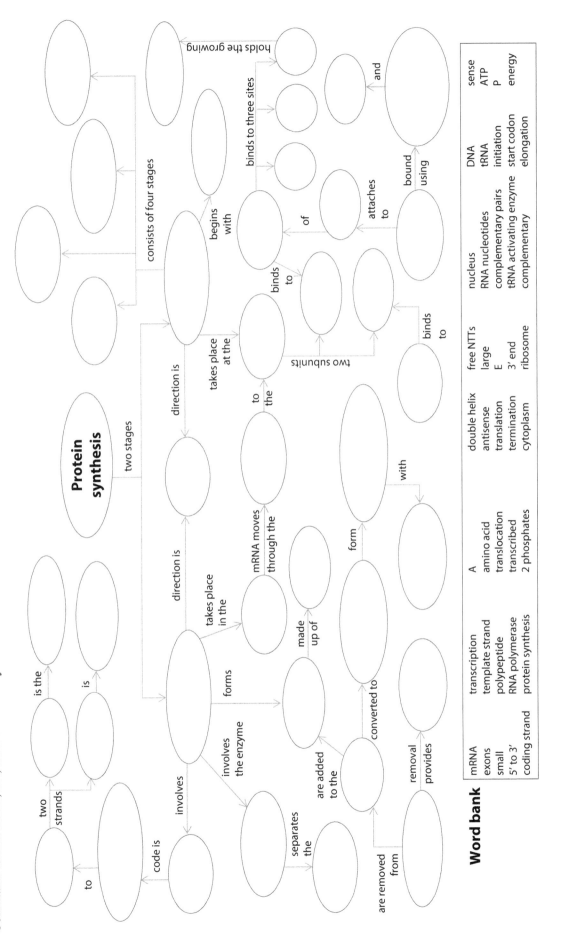

Word bank

mRNA	transcription	A	double helix	free NTTs	nucleus	DNA	sense
exons	template strand	amino acid	antisense	large	RNA nucleotides	tRNA	ATP
small	polypeptide	translocation	translation	E	complementary pairs	initiation	P
5' to 3'	RNA polymerase	transcribed	termination	3' end	tRNA activating enzyme	start codon	energy
coding strand	protein synthesis	2 phosphates	cytoplasm	ribosome	complementary	elongation	

7.3 Ribosome and tRNA Structure

Command terms: label, annotate, identify and analyse

The diagrams show the structure of a eukaryotic ribosome and transfer RNA molecule.

Label the parts indicated on both diagrams.

Eukaryotic Ribosome

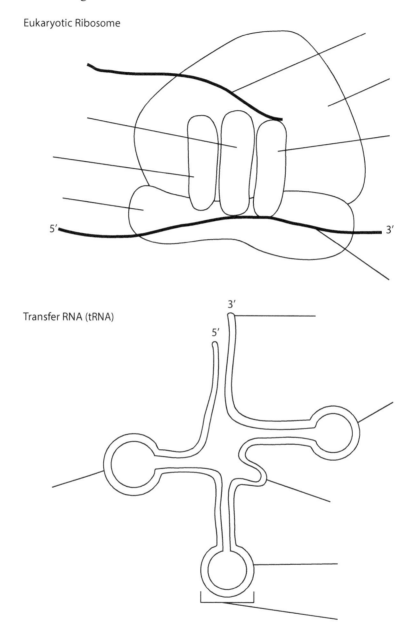

Transfer RNA (tRNA)

7.1–7.3 Function of Enzymes

Command terms: state and list

Complete the following table to describe the function of each enzyme involved in either DNA replication or protein synthesis.

Enzyme	Function
DNA replication	
Helicase	
DNA primase	
DNA polymerase I	
DNA polymerase III	
DNA ligase	
DNA gyrase	
Protein synthesis	
RNA polymerase	
tRNA-activating enzymes	

7.3 Protein Structure

Command terms: distinguish, compare, compare and contrast

Complete the table below to summarise the different levels of protein structure.

	Primary	Secondary	Tertiary	Quaternary
Structure	Linear			
Bonds				Hydrogen, disulphide bridges
Type		Fibrous	Globular	
Examples				
R-group interaction?	Yes		Yes	

METABOLISM, CELL RESPIRATION AND PHOTOSYNTHESIS (TOPIC 8) HIGHER LEVEL

8.1 Rates of Reaction

Command terms: measure, calculate, estimate, determine and predict

Reaction rate is the change determined in a reaction (e.g. the amount of products produced or amount of reactants used up) divided by the time interval for this change to take place:

$$\text{Rate} = \frac{\Delta \text{Measured value}}{\Delta \text{Time}}$$

To plot them on a graph, the rate of reaction is plotted on the Y axis and the other variable (independent variables such as temperature, substrate concentration and pH) is plotted on the X axis.

The graph below shows the effect of substrate concentration on the rate of reaction between an enzyme and its substrate and is an example of a graph showing reaction rate.

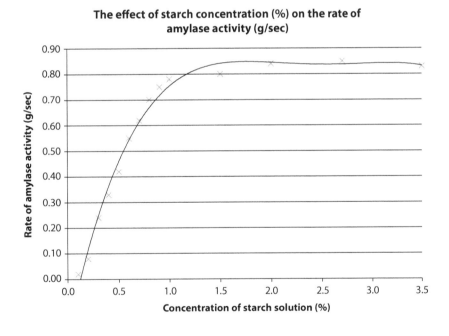

The effect of starch concentration (%) on the rate of amylase activity (g/sec)

8.1 Activation Energy

Command terms: draw, label, annotate, describe, distinguish and explain

1. Annotate the graph to show:

   a. The activation energy for both with and without enzymes

b. The reactants

c. The products

d. The overall energy released during the reaction

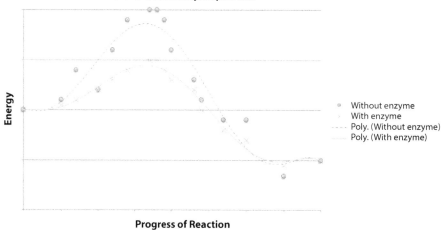

The activation energy required for a reaction with or without an enzyme present.

Energy (y-axis)

Progress of Reaction

2. Explain why the amount of reactants and products remains the same regardless of the presence of the enzyme.

_____

_____

3. Explain why the overall energy of the reaction does not change between conditions.

_____

_____

8.1 Competitive versus Non-competitive Inhibition

Command terms: outline, distinguish, compare, compare and contrast

Complete the table below to compare competitive and non-competitive enzyme inhibition.

	Competitive	Non-competitive
Structure of substrate and inhibitor		Dissimilar
Binding site	Active site	
Shape of enzyme		
Shape of active site		Distorted
Blockage of active site?	Yes	
Substrate binding		
Rate of reaction		Much lower than without inhibitor
Reversible?		
Example	Prontosil (antibacterial drug) inhibits the synthesis of folic acid in bacteria	

8.1 End-Product Inhibition

Command terms: list, state, annotate and identify

The flowchart represents an example of end-product inhibition, where threonine, an amino acid, is converted to isoleucine, which then inhibits the binding of threonine to threonine deaminase to prevent overproduction of isoleucine.

Substrate (threonine) binds to active site
on threonine deaminase (enzyme 1)

⬇

Intermediate A substance produced

⬇

Enzyme 2 converts Intermdiate A to Intermediate B

⬇

Enzyme 3 converts Intermediate B to Intermediate C

⬇

Enzyme 4 converts Intermediate C to Intermediate D

⬇

Enzyme 5 converts Intermdiate D to Isoleucine (end product)

⬇

When in excess, isoleucine binds to the allosteric site on
threonine deaminase (enzyme 1)

⬇

Active site on threonine deaminase no longer binds to threonine and no
subsequent reactions occur, preventing build up of all intermediate substances as
well as isoleucine.

8.2 Glycolysis

Command terms: state, describe, outline and explain

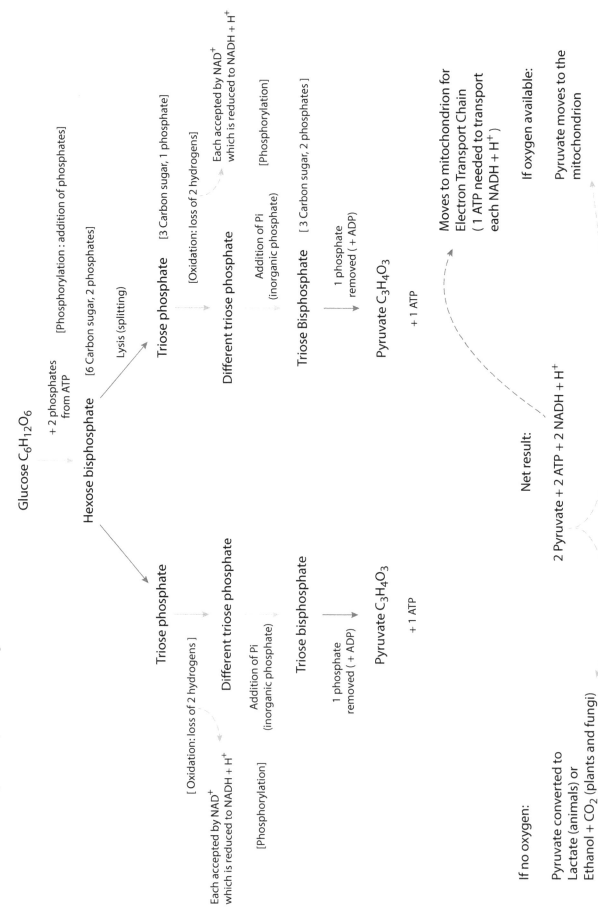

Glucose $C_6H_{12}O_6$

$+$ 2 phosphates from ATP   [Phosphorylation : addition of phosphates]

Hexose bisphosphate   [6 Carbon sugar, 2 phosphates]

Lysis (splitting)

Triose phosphate   [3 Carbon sugar, 1 phosphate]

[Oxidation: loss of 2 hydrogens]

Each accepted by NAD^+ which is reduced to $NADH + H^+$

[Phosphorylation]

Different triose phosphate

Addition of Pi (inorganic phosphate)

Triose Bisphosphate   [3 Carbon sugar, 2 phosphates]

1 phosphate removed ($+$ ADP)

Pyruvate $C_3H_4O_3$

$+$ 1 ATP

Moves to mitochondrion for Electron Transport Chain (1 ATP needed to transport each $NADH + H^+$)

If oxygen available:

Pyruvate moves to the mitochondrion

Net result:

2 Pyruvate $+$ 2 ATP $+$ 2 $NADH + H^+$

Triose phosphate

[Oxidation: loss of 2 hydrogens]

Different triose phosphate

Addition of Pi (inorganic phosphate)

[Phosphorylation]

Each accepted by NAD^+ which is reduced to $NADH + H^+$

Triose bisphosphate

1 phosphate removed ($+$ ADP)

Pyruvate $C_3H_4O_3$

$+$ 1 ATP

If no oxygen:

Pyruvate converted to Lactate (animals) or Ethanol $+$ CO_2 (plants and fungi)

8.2 The Link Reaction

Command terms: state, describe, outline and explain

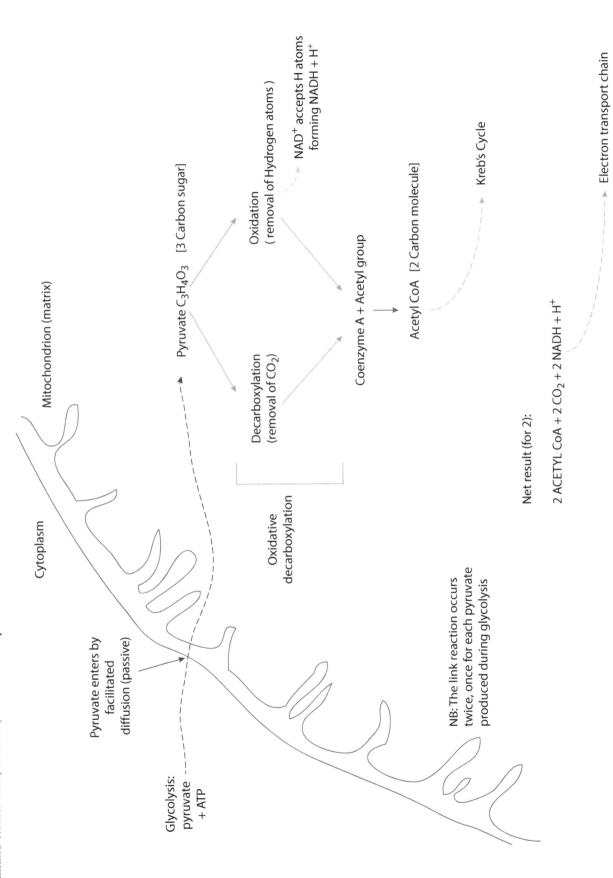

Cytoplasm

Mitochondrion (matrix)

Pyruvate enters by
facilitated
diffusion (passive)

Glycolysis:
pyruvate
+ ATP

Pyruvate $C_3H_4O_3$ [3 Carbon sugar]

Oxidation
(removal of Hydrogen atoms)

NAD^+ accepts H atoms
forming NADH + H^+

Decarboxylation
(removal of CO_2)

Oxidative
decarboxylation

Coenzyme A + Acetyl group

Acetyl CoA [2 Carbon molecule]

Kreb's Cycle

NB: The link reaction occurs
twice, once for each pyruvate
produced during glycolysis

Net result (for 2):

2 ACETYL CoA + 2 CO_2 + 2 NADH + H^+

Electron transport chain

8.2 Kreb's Cycle

Command terms: state, describe, outline and explain

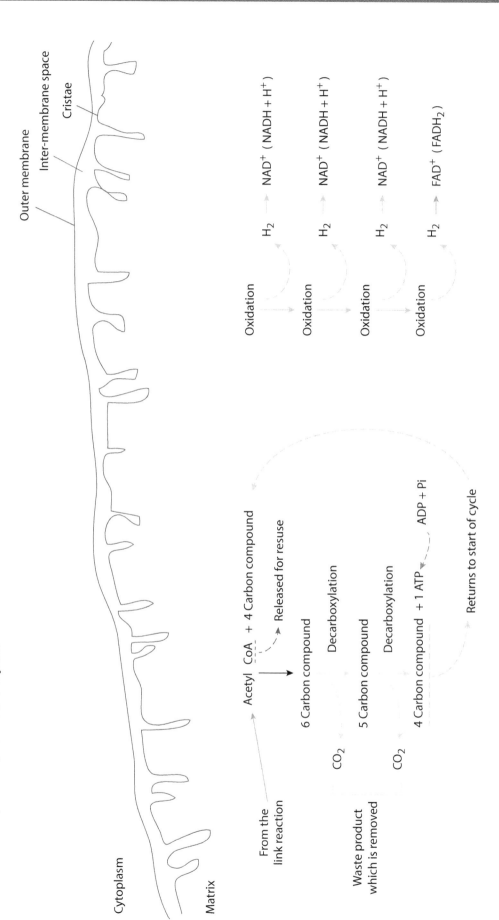

Outer membrane

Inter-membrane space

Cristae

Cytoplasm

Matrix

Oxidation $H_2 \longrightarrow NAD^+$ ($NADH + H^+$)

Oxidation $H_2 \longrightarrow NAD^+$ ($NADH + H^+$)

Oxidation $H_2 \longrightarrow NAD^+$ ($NADH + H^+$)

Oxidation $H_2 \longrightarrow FAD^+$ ($FADH_2$)

Electron transport chain

From the
link reaction

Acetyl CoA + 4 Carbon compound

Released for resuse

6 Carbon compound

Decarboxylation

CO_2

Waste product
which is removed

5 Carbon compound

Decarboxylation

CO_2

4 Carbon compound + 1 ATP

ADP + Pi

Returns to start of cycle

Net result (for 2 turns):

$4\,CO_2 + 6\,NADH + H^+ + 2\,FADH_2 + 2\,ATP$

NB: Kreb's Cycle turns twice for
each molecule of glucose (due
to the production of 2 pyruvates
and thus 2 acetyl CoA)

8.2 The Electron Transport Chain

Command terms: state, describe, outline and explain

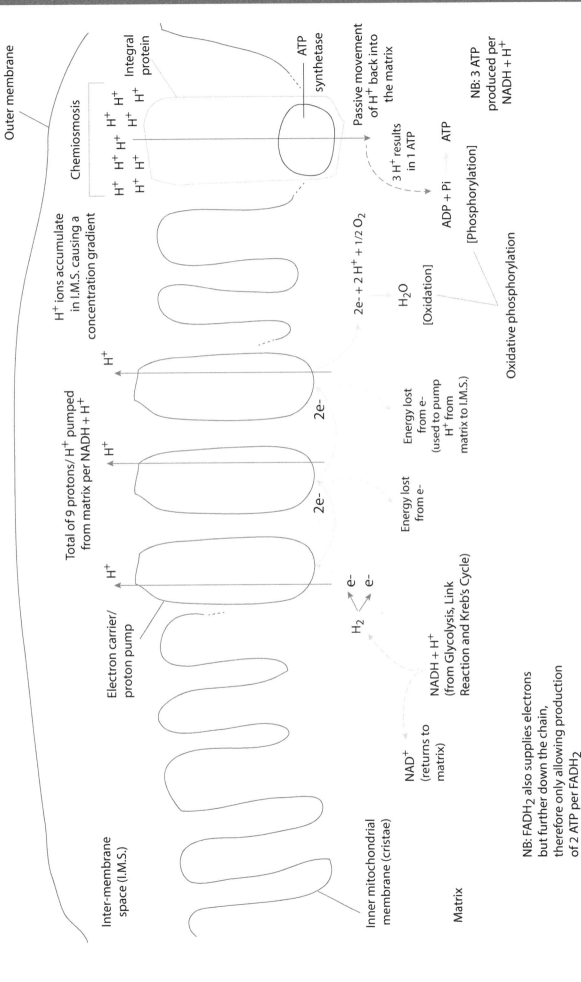

Outer membrane

Chemiosmosis

Integral protein

H^+ H^+ H^+ H^+
H^+ H^+
H^+ H^+

ATP synthetase

Passive movement of H^+ back into the matrix

3 H^+ results in 1 ATP

$ADP + Pi \longrightarrow ATP$

[Phosphorylation]

H^+ ions accumulate in I.M.S. causing a concentration gradient

$2e- + 2H^+ + 1/2 O_2$

H_2O

[Oxidation]

Oxidative phosphorylation

NB: 3 ATP produced per NADH + H^+

Total of 9 protons/ H^+ pumped from matrix per NADH + H^+

H^+

H^+

H^+

2e-

2e-

Energy lost from e-

Energy lost from e- (used to pump H^+ from matrix to I.M.S.)

Electron carrier/ proton pump

Inter-membrane space (I.M.S.)

H_2 $\nearrow$ e-
$\nwarrow$ e-

NADH + H^+ (from Glycolysis, Link Reaction and Kreb's Cycle)

NAD^+ (returns to matrix)

Inner mitochondrial membrane (cristae)

Matrix

NB: $FADH_2$ also supplies electrons but further down the chain, therefore only allowing production of 2 ATP per $FADH_2$

8.2 Mitochondrion

Command terms: draw, label, state, annotate and identify

For each labelled part, indicate the relationship between the structure and its function.

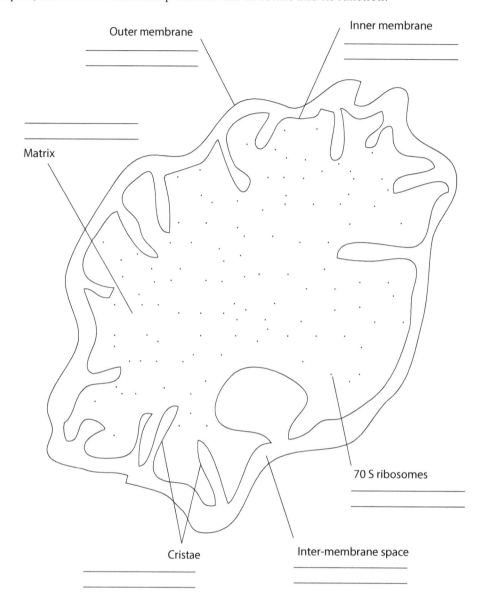

Outer membrane

Inner membrane

Matrix

70 S ribosomes

Cristae

Inter-membrane space

8.3 Light-Dependent Reactions

Command terms: state, describe, outline and explain

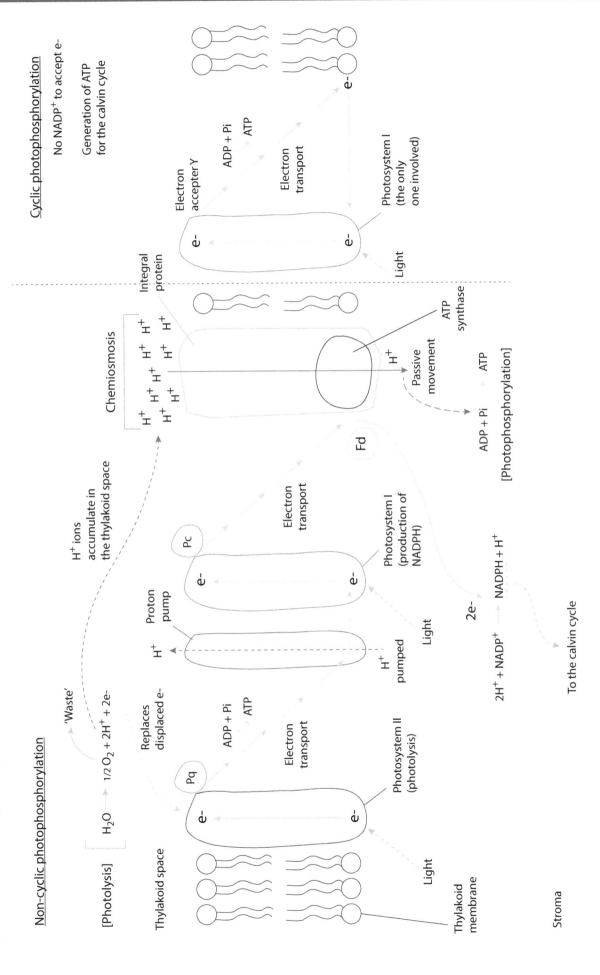

Non-cyclic photophosphorylation

Cyclic photophosphorylation

No NADP+ to accept e-

Generation of ATP for the calvin cycle

8.3 Light-Independent Reactions – The Calvin Cycle

Command terms: state, describe, outline and explain

Cycle turns once per CO_2

$3 \, CO_2 \longrightarrow$ 1 Triose phosphate

6 Triose phosphate = 18 turns

Stroma

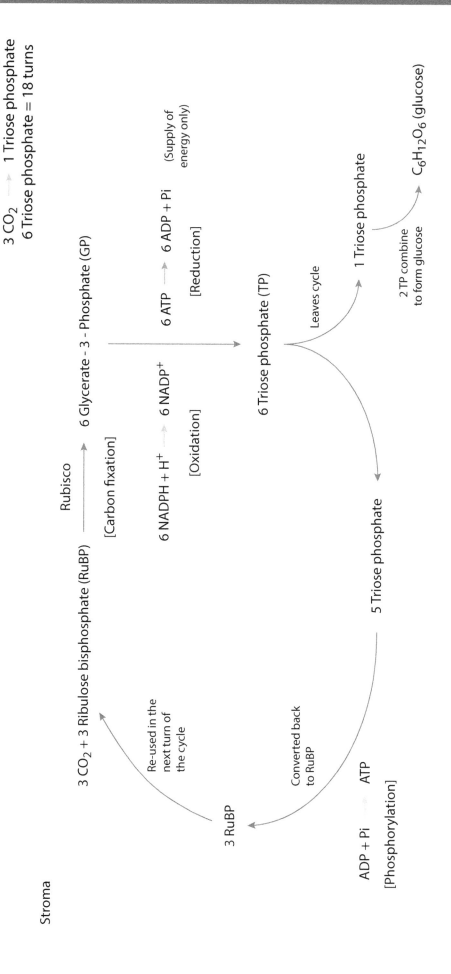

$3 \, CO_2 + 3$ Ribulose bisphosphate (RuBP)

Rubisco

6 Glycerate - 3 - Phosphate (GP)

[Carbon fixation]

$6 \, NADPH + H^+ \dashrightarrow 6 \, NADP^+$

[Oxidation]

$6 \, ATP \dashrightarrow 6 \, ADP + Pi$ (Supply of energy only)

[Reduction]

6 Triose phosphate (TP)

Leaves cycle

1 Triose phosphate

2 TP combine to form glucose

$C_6H_{12}O_6$ (glucose)

5 Triose phosphate

Converted back to RuBP

3 RuBP

Re-used in the next turn of the cycle

$ADP + Pi \dashrightarrow ATP$

[Phosphorylation]

8.3 Chloroplast

Command terms: draw, label, state, annotate and identify

For each labelled part, indicate the relationship between the structure and its function.

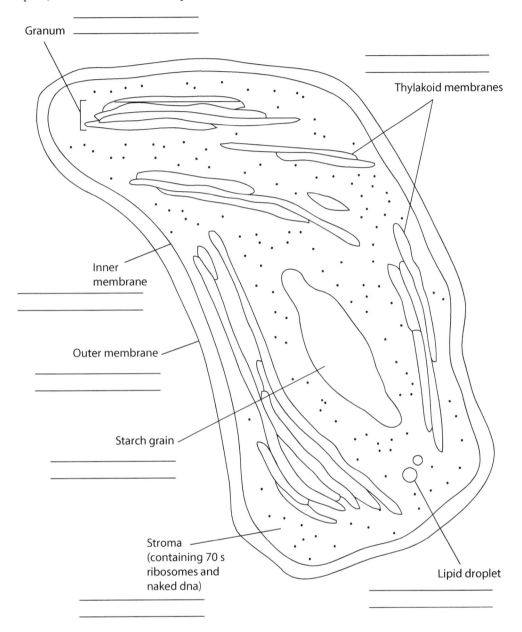

Granum _____

_____

_____

Thylakoid membranes

Inner membrane

_____

Outer membrane

_____

Starch grain

_____

Stroma
(containing 70 s
ribosomes and
naked dna)

_____

Lipid droplet

_____

9.1 Transpiration

Command terms: describe, identify and deduce

Number the stages below to summarise the process of transpiration in plants.

Water molecules evaporate   _____

Water is pulled upwards   _____

Water evaporates from the spongy mesophyll   _____

Xylem vessels fill with water   _____

Water diffuses out through the pores of the stomata   _____

Water moves by osmosis into the mesophyll cells from the vascular bundles   _____

9.1–9.3 Plant Tissues

Command terms: list, state and identify

State the main function of each of the plant tissues or structures given below.

Indicate with ★ tissue/s involved with <u>photosynthesis</u>
Indicate with ✗ tissue/s involved with <u>transpiration</u>
Indicate with ■ tissue/s involved with <u>growth</u>
Indicate with ♯ tissue/s involved with <u>reproduction</u>

Plant structure/tissue	Description of function	Symbol for function
Palisade mesophyll		
Shoot		
Anther		
Stoma		
Upper/lower cuticle		
Xylem		
Phloem		
Root		
Epidermis		
Pith		

Plant structure/tissue	Description of function	Symbol for function
Testa		
Meristem		
Guard cell		
Stigma		
Cambium		
Ovary		
Spongy mesophyll		
Cotyledon		
Parenchyma cells		
Upper/lower epithelium		

9.1 Plant Adaptations

Command terms: list, state and describe

This table below provides a summary of the adaptations of plants to the extreme environments brought on by dry or saline conditions.

Dry conditions (Xerophytes)	Saline conditions (Halophytes)
Rolled leaves	Germination at times of low salinity such as during high rainfall
Reduced leaves	Germination on the parent plant rather than shedding (no dormancy)
Leaves reduced to spines	Excretion of excess salts through leaves
Thickened waxy cuticle	Storage of excess salts in leaves that later drop off
Low growth form	Thickened waxy cuticle
Reduced number of stomata	Sunken stomata
Stomata in pits surrounded by hairs	Long roots
Deep roots	Stem can take over photosynthesis when leaves are shed or reduced
Water shortage tissue	Water storage in leaves
Crassulacean acid metabolism (CAM) and C4 physiology	Reduced leaves
	Selective permeability of roots to salts

9.1, 9.2 Plant Transport – Concept Map

Command terms: define, list, state and identify

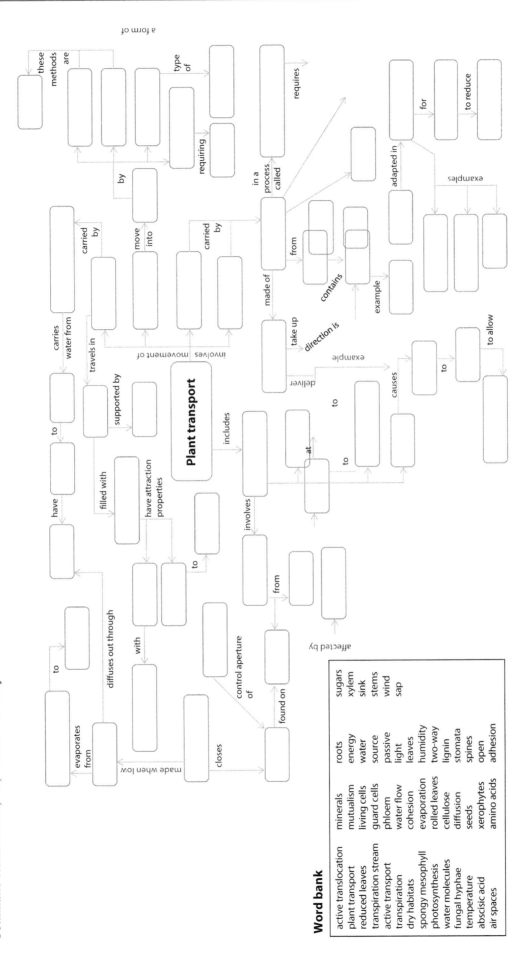

Word bank

active translocation	minerals	sugars
plant transport	mutualism	xylem
reduced leaves	living cells	sink
transpiration stream	guard cells	stems
active transport	phloem	wind
transpiration	water flow	sap
dry habitats	cohesion	
spongy mesophyll	evaporation	humidity
photosynthesis	rolled leaves	two-way
water molecules	cellulose	lignin
fungal hyphae	diffusion	stomata
temperature	seeds	spines
abscisic acid	xerophytes	open
air spaces	amino acids	adhesion

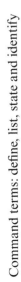

9.2 Plant Transport – Source to Sink

Command terms: label and annotate

The diagram shows the plant tissues (xylem and phloem) involved in the transport of nutrients and water around the plant.

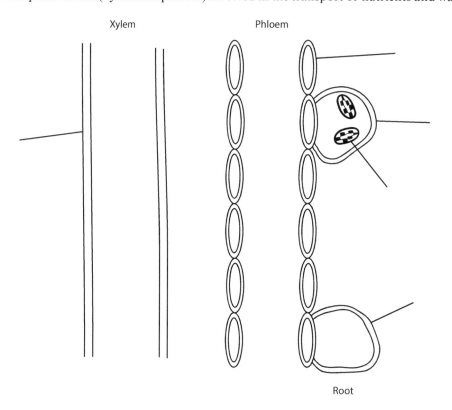

Xylem   Phloem

Root

1. Label the indicated parts.
2. Draw arrows within the xylem tissue and phloem tissue to show direction of movement.
3. Label these arrows as either 'transpiration pull' or 'translocation'.
4. Indicate on the diagram where the following processes occur:

   ▲ transport of sugars
   ● transport of amino acids
   ★ transport of water

5. Give four examples of tissues acting as a source and four examples of tissues acting as a sink.

   _____   _____

   _____   _____

   _____   _____

   _____   _____

9.1–9.2 Plant Transport Mechanisms

Command terms: state, describe and outline

Classify the following examples of plant transport as occurring by (a) osmosis, (b) simple diffusion, (c) facilitated diffusion or (d) active transport.

Gas exchange within the leaf   _____

Uptake of mineral ions in the roots   _____

Absorption of water in the roots   _____

Loading of organic compounds into phloem sieve tubes   _____

Water uptake at the source   _____

Movement of minerals from soil to root   _____

Movement of H^+ ions back into root cells   _____

Passage of water through stomatal pores   _____

Movement of water into mesophyll cells   _____

Sugars taken up in a source area   _____

Removal of sugars and organic compounds at a sink area   _____

Movement of water from cells at the sink   _____

9.2 Xylem and Phloem

Command terms: label, annotate and identify

The diagrams below show cross sections through a dicotyledonous plant root and stem.

1. In each of the diagrams, identify the xylem tissue and the phloem tissue and indicate these.
2. Label the remainder of the parts indicated on the diagrams.

Dicotyledonous Stem

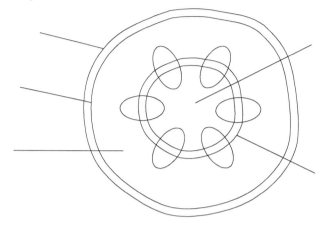

Dicotyledonous Root

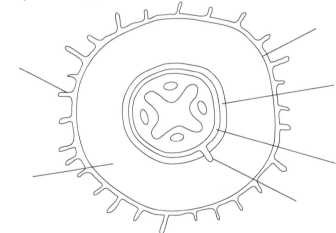

9.3 Plant Hormones

Command terms: state, describe, outline and explain

Fill in the blanks to complete the sentences below that outline the role of hormones in plant growth.

Plant hormones control growth in the _____.

Both the _____ and the _____ of the growth are controlled.

_____ is a growth movement in response to light coming from a specific direction.

A _____ phototropism is when a plant grows _____ the light.

A _____ phototropism is when a plant grows _____ the light.

An example of a hormone that influences cell growth rates is called _____.

_____ is produced by the _____ of a growing plant and is then transported down the _____ _____.

Auxin changes the pattern of _____ and stimulates _____.

Auxin _____ set up concentration gradients of auxin in plant tissue.

If greater _____ is detected on one side of the stem, the auxin accumulates on the _____ side of the plant.

Higher concentrations of _____ cause greater _____ on this side of the plant and the stem grows towards the source of the _____.

This means plants can obtain the most _____ and are thus able to _____ at a greater rate.

9.4 Plant Reproduction – Concept Map

Command terms: define, list, state and identify

Plant reproduction

Word bank

germination	seeds	dormant	aerobic respiration	flowers	light
seed dispersal	testa	far-red light	photoperiodicity	water	protein
angiospermophytes	anther	competition	fertilisation	carpal	red light
short nights	oxygen	long-day	short-day	starch	enzymes
pollination	glucose	proteins	night length	organism	
plant reproduction	730nm	cotyledons	phytochrome	P$_R$	
				amylase	gametes
				stable	long nights
				lipids	gibberellin
				receptor	660nm
				P$_{FR}$	pollen
				temperature	maltose

9.4 Animal-Pollinated Flower

Command terms: draw, label and identify

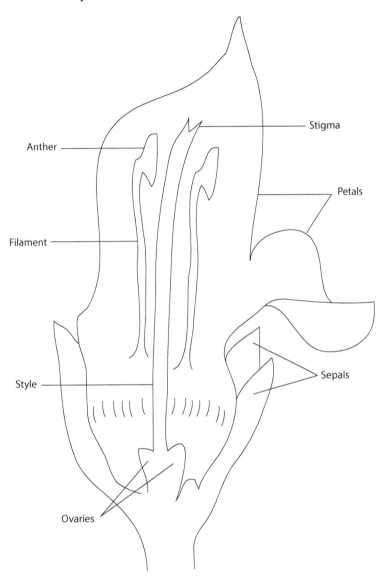

9.4 Internal Seed Structure

Command terms: draw, label, annotate and identify

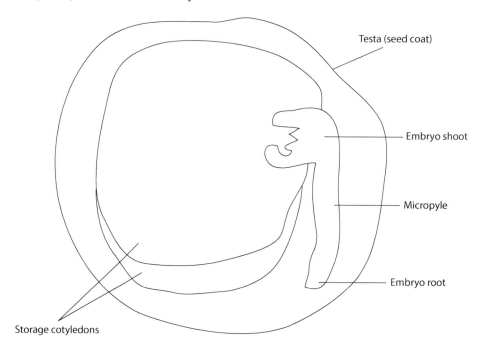

Testa (seed coat)

Embryo shoot

Micropyle

Embryo root

Storage cotyledons

For each of the labelled parts, describe their involvement in germination and therefore
Plant reproduction

Testa   _____

Embryo shoot   _____

Embryo root   _____

Micropyle   _____

Storage cotyledons _____

9.4 Stages of Seed Germination

Command terms: describe, identify and deduce

Number the stages below to summarise the process of seed germination in plants.

Seed shoot appears   _____

Amylase breaks down starch to maltose   _____

Stored proteins and lipids are broken down   _____

Seed coat ruptures   _____

Seed lays dormant   _____

Amino acids make new proteins for the embryo _____

Amylase is produced   _____

Suitable temperature conditions met   _____

Water is absorbed   _____

Cellulose used to make cells walls for new cells _____

Glucose used in aerobic respiration   _____

Enzymes are activated   _____

Fatty acids and glycerol used in membranes and for energy _____

Seed root appears   _____

Seed disperses from parent plant   _____

Gibberellin is produced in the cotyledons   _____

Maltose is converted to glucose   _____

CHAPTER

10 GENETICS AND EVOLUTION (TOPIC 10) HIGHER LEVEL

10.1 Chromosomes

Command terms: define, label, state, annotate, describe and identify

Describe the terms shown below and annotate or label the diagrams where possible.

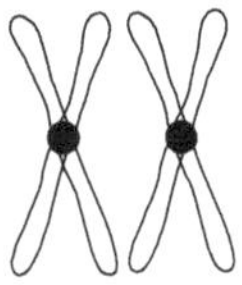

Homologous chromosomes

___

___

___

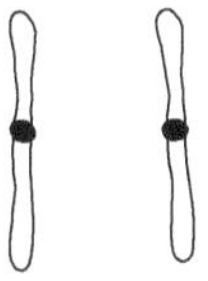

Sister chromatids

___

___

___

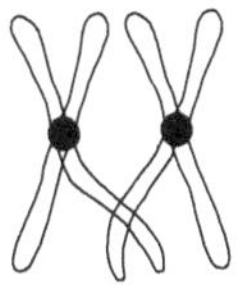

Chiasmata

___

___

___

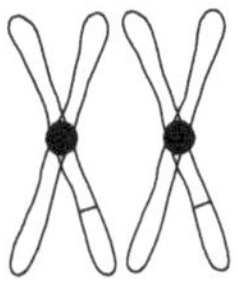

Locus

___

___

___

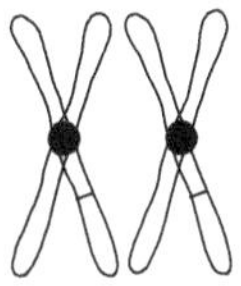

Allele and gene

___

___

___

10.1 Meiosis

Command terms: state, identify, annotate and outline

Organise the following descriptions of the behaviour of chromosomes during the various stages of meiosis into the appropriate column below.

Centromeres separate
Chromosomes condense
Homologous pairs move to equator of the cell
Chromosomes uncoil
Chromatids move to opposite poles
Spindle forms
Nuclear envelope forms
Chromosomes move to equator
Cell has diploid number of chromosomes
Homologous pairs are separated
Chromosomes become chromatin
Spindle disappears
Cell has haploid number of chromosomes
Homologous chromosomes pair
Chiasmata form
Orientation of chromosomes is random and independent
Chromosomes condense and become visible
New spindle forms at right angles to previous spindle
One chromosome of each pair moves to each pole
Chromosomes arrive at poles
Spindle disappears
Crossing over occurs
Chromatids reach opposite poles

Prophase I	Metaphase I	Anaphase I	Telophase I

Prophase II	Metaphase II	Anaphase II	Telophase II

10.2 Continuous versus Discrete Variation

Command terms: define, list and state

Complete the table below to state the definition of continuous and discrete variation and provide examples for each type.

	Definition	Examples
Continuous variation		
Discrete variation		

10.2 Dihybrid Punnett Squares

Command terms: identify, analyse, construct, deduce, determine and predict

A dihybrid cross, like a monohybrid cross, can be investigated using a Punnett square. A dihybrid cross involves the investigation of two traits together.

Steps to complete a dihybrid Punnett square:
1. Deduce the genotypes of the parents for the two traits investigated.
2. Identify all possible gametes the parents can produce, given their genotypes for both traits.
3. Construct a Punnett square, separating out the possible gametes for the male (across the top of the square) and female (along the side of the square).
4. Complete the Punnett square by adding in the genotypes for the offspring into each box.
5. Determine the ratio of genotypes amongst the offspring.
6. Determine the ratio of phenotypes for the offspring based on their genotypes.

Example:
Wing shape and colour in flies can be investigated. Wing shape can be long (A) or wide (a), whilst colour can be brown (B) or black (b). Two flies, heterozygous for both traits were crossed.
1. The genotype of each parent: AaBb
2. The possible gametes of each parent: AB, Ab, aB, ab
3./4. Punnett square:

	AB	Ab	aB	ab
AB	AABB	AABb	AaBB	AaBb
Ab	AABb	AAbb	AaBb	Aabb
aB	AaBB	AaBb	aaBB	aaBb
ab	AaBb	Aabb	aaBb	aabb

5. The ratio of offspring genotypes: 1 AABB: 2 AABb: 2 AaBB: 4 AaBb: 1 AAbb: 2 Aabb: 1 aaBB: 2 aaBb: 1 aabb
6. The ratio of offspring phenotypes (shaded): 9 long brown: 3 long black: 3 wide brown: 1 wide black

10.2 Recombinants

Command terms: identify, deduce and determine

Recombinants are those offspring that possess entirely different combinations of alleles from either of their parents. This occurs as a result of crossing over during meiosis.

We will use the example from the previous section, 10.2 Dihybrid Punnett Squares.

The genotype of each parent: AaBb

All possible genotypes of the offspring: AABB, AABb, AaBB, AaBb, AAbb, Aabb, aaBB, aaBb, aabb.

Determine which of the offspring are recombinants

_____

_____

10.2 Chi-Squared Test for Dihybrid Crosses

Command terms: calculate, deduce and determine

The chi-squared test discussed in Section 4.1 can also be applied to the analysis of dihybrid crosses. In this case, the chi-squared test is a measure of the statistical significance of the difference between observed and expected phenotypes. There are two hypotheses: the traits assort independently (H_0) or the traits do not assort independently (H_1). The ratio of expected phenotype frequencies is 9:3:3:1 (dominant for both traits: dominant for one trait, recessive for the other: recessive for one trait, dominant for the other: recessive for both traits).

The formula for the chi-squared test is:

$$X^2 = \Sigma \frac{(O-E)^2}{E}$$

where

X^2 = the test statistic

O = the observed values

E = the expected values

Σ = sum of

Gregor Mendel, in his studies of pea plants, investigated traits such as seed shape and seed colour. The contingency table below shows the observed phenotypic frequencies of these two traits and the expected frequencies based on the size of the F_1 generation. A sample of 200 offspring from the F_1 generation is shown.

The traits studied are:
  Seed colour: yellow (dominant) or green (recessive)
  Seed shape: round (dominant) or wrinkled (recessive)

	Yellow round	Yellow wrinkled	Green round	Green wrinkled	Total
Observed	116	39	36	9	200
Expected	(9/16) × 200 = 112.5	(3/16) × 200 = 37.5	(3/16) × 200 = 37.5	(1/16) × 200 = 12.5	200

1. Calculate the number of degrees of freedom.

2. Find the critical value for chi-squared at a significance level of 0.05 (5 per cent) using the chi-squared distribution table below.

Percentage Points of the Chi-Square Distribution

Degrees of freedom	Probability of a larger value x^2								
	0.99	0.95	0.90	0.75	0.50	0.25	0.10	0.05	0.01
1	0.000	0.004	0.016	0.102	0.455	1.32	2.71	3.84	6.63
2	0.020	0.103	0.211	0.575	1.386	2.77	4.61	5.99	9.21
3	0.115	0.352	0.584	1.212	2.366	4.11	6.25	7.81	11.34
4	0.297	0.711	1.064	1.923	3.357	5.39	7.78	9.49	13.28
5	0.554	1.145	1.610	2.675	4.351	6.63	9.24	11.07	15.09
6	0.872	1.635	2.204	3.455	5.348	7.84	10.64	12.59	16.81
7	1.239	2.167	2.833	4.255	6.346	9.04	12.02	14.07	18.48
8	1.647	2.733	3.490	5.071	7.344	10.22	13.36	15.51	20.09
9	2.088	3.325	4.168	5.899	8.343	11.39	14.68	16.92	21.67
10	2.558	3.940	4.865	6.737	9.342	12.55	15.99	18.31	23.21
11	3.053	4.575	5.578	7.584	10.341	13.70	17.28	19.68	24.72
12	3.571	5.226	6.304	8.438	11.340	14.85	18.55	21.03	26.22
13	4.107	5.892	7.042	9.299	12.340	15.98	19.81	22.36	27.69
14	4.660	6.571	7.790	10.165	13.339	17.12	21.06	23.68	29.14
15	5.229	7.261	8.547	11.037	14.339	18.25	22.31	25.00	30.58
16	5.812	7.962	9.312	11.912	15.338	19.37	23.54	26.30	32.00
17	6.408	8.672	10.085	12.792	16.338	20.49	24.77	27.59	33.41
18	7.015	9.390	10.865	13.675	17.338	21.60	25.99	28.87	34.80
19	7.633	10.117	11.651	14.562	18.338	22.72	27.20	30.14	36.19
20	8.260	10.851	12.443	15.452	19.337	23.83	28.41	31.41	37.57
22	9.542	12.338	14.041	17.240	21.337	26.04	30.81	33.92	40.29
24	10.856	13.848	15.659	19.037	23.337	28.24	33.20	36.42	42.98
26	12.198	15.379	17.292	20.843	25.336	30.43	35.56	38.89	45.64
28	13.565	16.928	18.939	22.657	27.336	32.62	37.92	41.34	48.28
30	14.953	18.493	20.599	24.478	29.336	34.80	40.26	43.77	50.89
40	22.164	26.509	29.051	33.660	39.335	45.62	51.80	55.76	63.69
50	27.707	34.764	37.689	42.942	49.335	56.33	63.17	67.50	76.15
60	37.485	43.188	46.459	52.294	59.335	66.98	74.40	79.08	88.38

Source: Plant and Soil Sciences eLibrary, University of Nebraska-Lincoln, USA, 2015, http://passel.unl.edu/pages/informationmodule.php?idinformationmod
ule=1130447119, used with permission.

3. Calculate chi-squared using the formula.

_____

_____

_____

_____

4. Compare the calculated value for chi-squared with the critical value and determine if the two traits are linked or un-linked. Explain your answer.

_____

_____

_____

_____

_____

_____

_____

_____

_____

10.3 Reproductive Isolation

Command terms: list, state and identify

Determine which of the following statements about the different categories of reproductive isolation are 'true' or 'false'.

	True/False
Reproductive isolation can be temporal, behavioural or geographic	_____
Reproductive isolation always leads to speciation	_____
Only geographic isolation leads to speciation	_____
Allopatric speciation occurs as a result of geographic isolation	_____
Sympatric speciation occurs as a result of behavioural isolation	_____
Separation of a populations gene pool cannot occur within the same geographic location	_____
Mating rituals are an example of behavioural isolation	_____
Reproductive isolation is essential to the evolution of new species	_____
Rivers, lakes and mountain ranges are examples of geographic barriers	_____
Seasonal fluctuations can result in reproductive isolation	_____
If a barrier is removed, the two populations of species may interbreed again	_____

10.3 Selection

Command terms: define, list and state

Complete the following table to state the definition for directional selection, stabilising selection and disruptive selection and provide examples for each type.

	Definition	Examples
Directional selection		
Stabilising selection		
Disruptive selection		

CHAPTER
11
ANIMAL PHYSIOLOGY (TOPIC 11) HIGHER LEVEL

11.1 Antibody Production

Command terms: state, describe, outline and explain

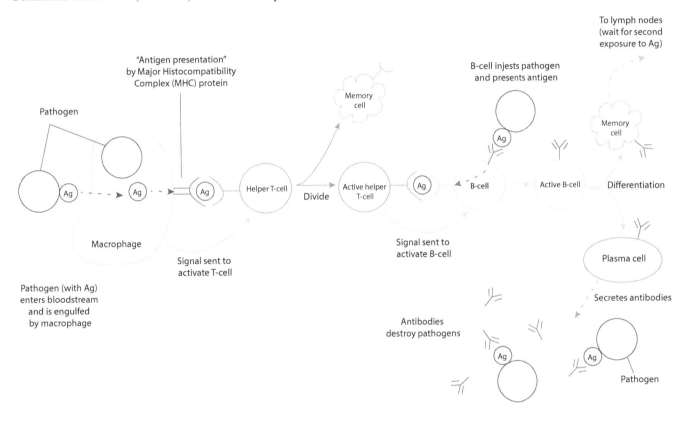

11.1 The Human Immune System – Concept Map

Command terms: define, list, state and identify

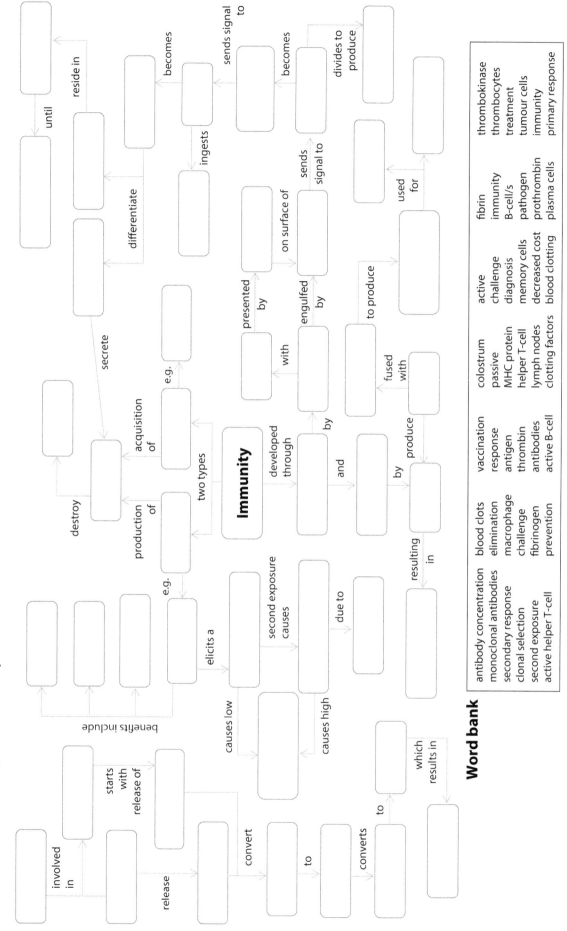

Word bank

antibody concentration	blood clots	vaccination	colostrum	active	thrombokinase
monoclonal antibodies	elimination	response	passive	challenge	thrombocytes
secondary response	macrophage	antigen	MHC protein	diagnosis	treatment
clonal selection	challenge	thrombin	helper T-cell	memory cells	tumour cells
second exposure	fibrinogen	antibodies	lymph nodes	decreased cost	immunity
active helper T-cell	prevention	active B-cell	clotting factors	blood clotting	primary response

fibrin		
immunity		
B-cell/s		
pathogen		
prothrombin		
plasma cells		

11.2 Movement

Command terms: define and state

Link the following terms with their meanings.

Bone	Provide the force needed for movement once contracted
Muscle	Allows for movement to occur in certain directions and helps to reduce friction of this movement
Tendon	Provides rigid structure and support
Ligament	Absorbs shock and distributes load
Synovial joint	Attach muscle to bone
Nerve	Maintains position of bones and prevents dislocation
Cartilage	Stimulate muscular contractions

11.2 Human Elbow

Command terms: draw, label and annotate

The diagram shows the structure of the human elbow.

1. Label the parts indicated on the diagram.
2. Colour code the bones and muscles.
3. Add annotations to outline the role of the named structures in the functioning of the elbow.

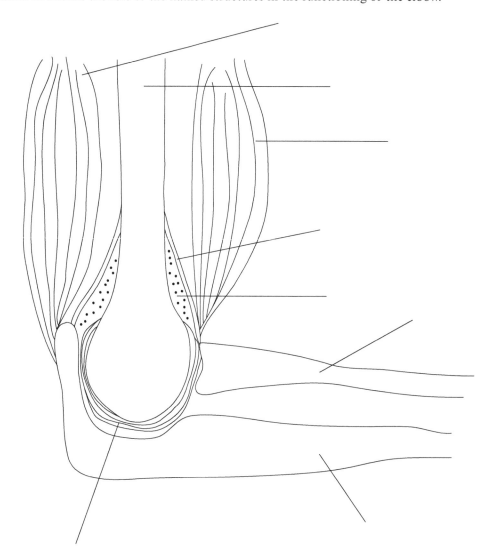

11.2 Sarcomere

Command terms: draw, label, annotate and identify

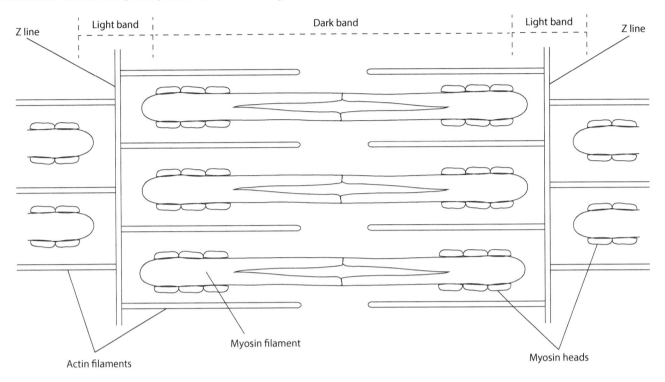

11.2 Structure of Skeletal Muscle

Command terms: describe, identify and deduce

Number the structures that form striated muscle fibres in the correct order from smallest to largest.

Myofibril   _____

Troponin   _____

Dark band   _____

Muscle   _____

Actin   _____

Z line   _____

Sarcomere   _____

Tropomyosin   _____

Myosin   _____

Light band   _____

Muscle fibre   _____

Myosin head   _____

11.3 The Human Kidney

Command terms: draw, label, annotate and identify

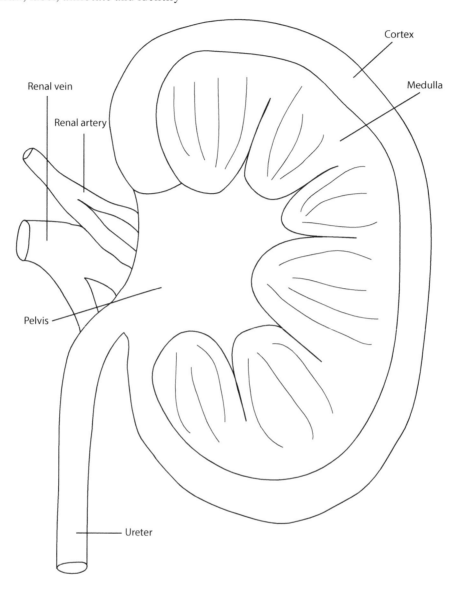

11.3 The Nephron

Command terms: draw, label, annotate and identify

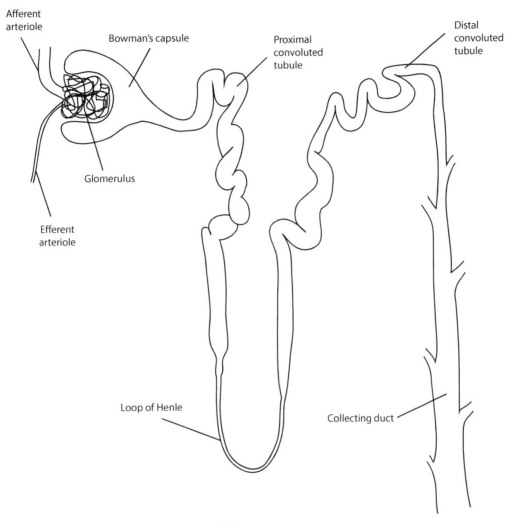

Indicate on the diagram the place/s that the following processes occur:

✳ Ultrafiltration
✚ Reabsorption
✓ Osmoregulation

Indicate the location of highest concentration of:

● Blood plasma
■ Blood cells
▲ Urea
◆ Glucose
★ Proteins

Indicate the location of:

✖ Osmosis
▬ Active transport

11.3 Length of Loop of Henlé in Animals

Command terms: calculate, estimate, identify, analyse, compare and contrast, deduce, determine, evaluate, predict and suggest

The table below shows a comparison in a number of species of the Rodent order of kidney size and urine concentration. It has been shown that habitat and the need for water conservation relate to the length of the Loop of Henlé and therefore the maximum urine concentration. *Homo sapiens* have been included as a means of comparison.

Animal	Common name	Mean live weight (kg)	Habitat	Mean kidney size (mm)	Mean length of Loop of Henlé (mm)	Urine concentration (mOsmol/L)
Homo sapiens	Human	70.00	Urban	64.0	21.0	1,200
Jaculus orientalis	Jerboa	0.06	Desert	4.5	1.4	6,500
Notomys alexis everardensis	Spinifex hopping mouse	0.04	Desert	14.0	14.0	9,000
Rattus norvegicus	Brown rat	0.35	Plains	14.0	4.0	2,900
Dipodomys merriami	Kangaroo rat	0.13	Desert	5.9	1.5	5,500
Psammomys obesus	Sand rat	0.04	Desert	13.0	13.0	6,000

1. Discuss the theory that the length of the Loop of Henlé in rodents is related to the maximum urine concentration.

   _____

   _____

   _____

   _____

2. Compare the mean kidney size and mean length of Loop of Henlé with urine concentration in *N. alexis everardensis* and *R. norvegicus.*

   _____

   _____

   _____

3. Why should rodents who live in desert ecosystems have a Loop of Henlé longer in length than those in less arid ecosystems?

   _____

   _____

   _____

4. How does kidney size relate to the length of the Loop of Henlé?

   _____

   _____

   _____

11.3 Content of Blood Plasma

Command terms: state, distinguish, estimate and compare

The table below summarises the concentration of various components in blood plasma, glomerular filtrate and urine.

Substance	Concentration in blood plasma (g 100 mL⁻¹)	Concentration in glomerular filtrate (g 100 mL⁻¹)	Concentration in urine (g 100 mL⁻¹)
Blood cells	50.00	0.00	0.00
Proteins	0.75	0.00	0.00
Glucose	0.10	0.10	0.00
Urea	0.03	0.03	2.00
Na⁺	0.32	0.32	0.60
Cl−	0.37	0.37	0.60
Water	90.00	90.00	95.00

11.4 Mature Human Egg

Command terms: label, annotate, identify and outline

Finish labelling the diagram of a mature human egg and add annotations to indicate the function of the named structures.

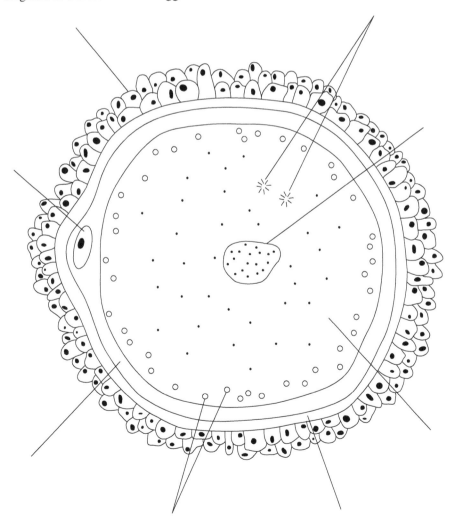

11.4 Mature Human Sperm

Command terms: draw, label, state, annotate, identify and outline

Finish labelling the diagram of the mature sperm and add annotations to outline the functions of the named structures.

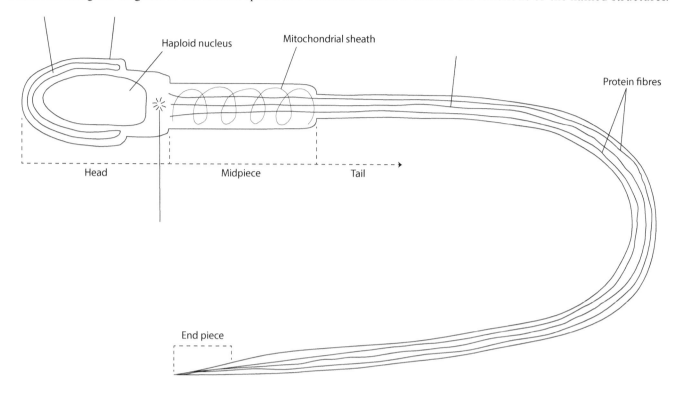

11.4 Structure and Function of the Placenta

Command terms: state and outline

Complete the table to describe the function of various parts of the human placenta.

Feature	Function
Muscular wall	
Chorionic villi	
Inter-villous spaces	
Capillaries	
Progesterone	
Oestrogen	

11.4 Fertilisation

Command terms: draw, label, annotate and outline

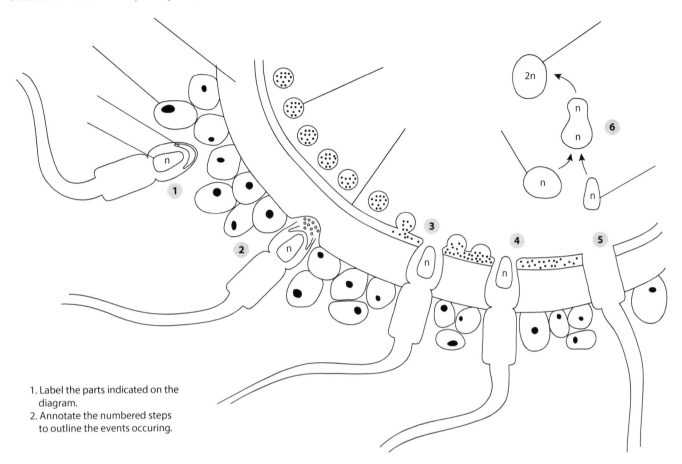

1. Label the parts indicated on the
   diagram.
2. Annotate the numbered steps
   to outline the events occuring.

11.4 Ovary and Seminiferous Tubule

Command terms: label, annotate and identify

The diagrams show the female ovary and male seminiferous tubule. Label the parts indicated and annotate the diagrams to show the stages of gametogenesis.

Ovary

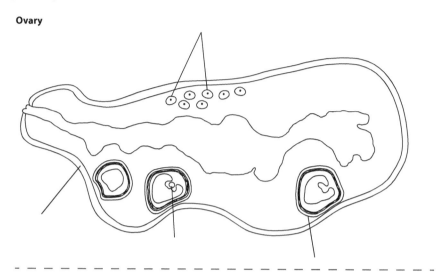

Seminiferous tubule

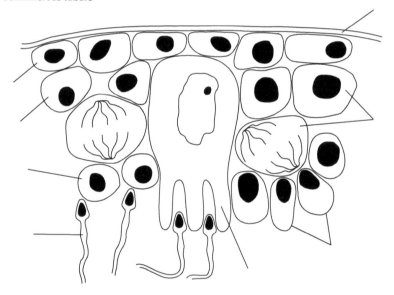

11.4 Spermatogenesis versus Oogenesis

Command terms: distinguish, compare, compare and contrast

Complete the table to compare and contrast the processes of spermatogenesis and oogenesis in humans.

	Spermatogenesis	Oogenesis
Does mitosis occur?		
Do meiosis I and II occur?		
Location		
Initial cell?		
Product of mitosis		
Product of meiosis I		
Product of meiosis II		
Product of differentiation		
Number of gametes		
Occurs		
Commences		
Stops		
Release of gametes		
Controlling hormones		

NEUROBIOLOGY AND BEHAVIOUR (OPTION A)

A.1 Xenopus Neurulation

Command terms: label, state, annotate, distinguish and identify

The diagrams show the embryonic tissues in Xenopus during neurulation.

1. Place the diagrams in the correct order of development by adding a number to each diagram.
2. Label the parts indicated on the diagrams.
3. Annotate the diagrams to outline events occurring in each stage.

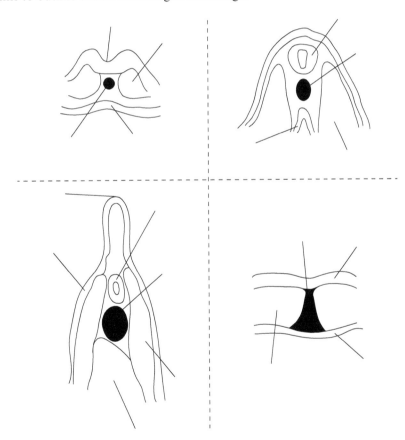

A.2 Body Size and Brain Size Comparison

Command terms: compare, compare and contrast, deduce and predict

The table below contains data on the body and brain mass for a number of animals.

Species name	Common name	Body mass (kg)	Brain mass (g)	Ratio of body to brain mass
Homo sapiens	Human	75.0	1,400.0	
Pan troglodytes	Chimpanzee	45.0	398.0	
Balaenoptera musculus	Blue whale	58,060.0	6,800.0	
Tursiops truncatus	Bottlenose dolphin	119.9	1,535.0	
Loxodonta africana	African elephant	5,140.0	4,783.0	
Giraffa camelopardalis	Giraffe	1,192.0	680.0	
Panthera leo	Lion	143.0	240.0	
Hippopotamus amphibius	Hippopotamus	1,350.0	732.0	
Equus ferus caballus	Horse	445.0	532.0	
Ornithorhynchus anatinus	Platypus	1.8	9.0	
Felis catus	Domestic cat	3.9	30.0	
Rattus rattus	Black rat	0.7	2.5	
Coturnix coturnix	Common quail	0.3	0.9	
Passer domesticus	House sparrow	0.4	1.0	
Carassius auratus auratus	Goldfish	0.6	0.1	

Use the data in the table above to answer the following questions.

1.   Calculate the ratio for the body mass and brain mass to complete the table.

2.   Determine the relationship between brain mass and body mass.

_____

_____

3.   Suggest a reason for the difference in the ratio of brain mass to body mass from one species to another.

_____

_____

_____

_____

4. (a) Compare the ratio of brain to body mass in humans and in chimpanzees.

_____

_____

_____

(b) Discuss in terms of the evolutionary relationship between the two species.

_____

_____

_____

A.2 The Human Brain

Command terms: draw, label, state, annotate and identify

Match the part of the brain with its function and label the parts of the brain on the diagram.

Hypothalamus	Complex thought processes such as memory, learning, problem solving, emotion
Cerebellum	Hormone production, control of hormone secretion, maintenance of homeostasis
Pituitary gland	Coordination of muscle movement and balance
Medulla oblongata	Stores and secretes hormones
Cerebral hemispheres	Autonomic functions such as breathing, swallowing, digestion, heart rate

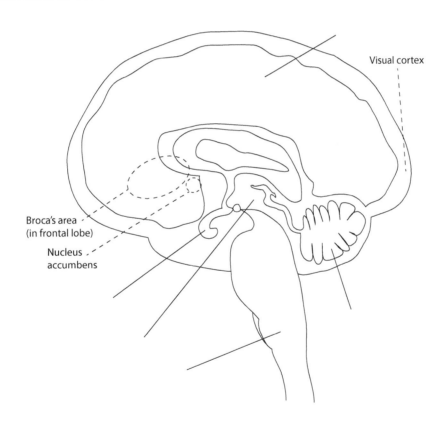

A.2 Human Brain

Command terms: annotate, identify and deduce

The images below shown a human brain and were obtained using Magnetic Resonance Imaging (MRI). Identify and label the parts shown, then annotate the images to include the function of the labelled parts.

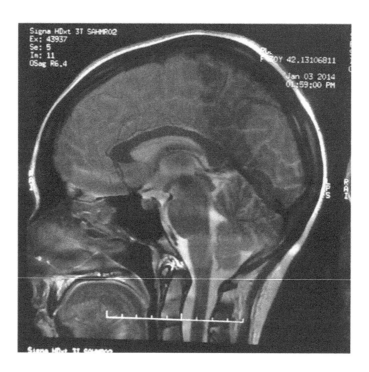

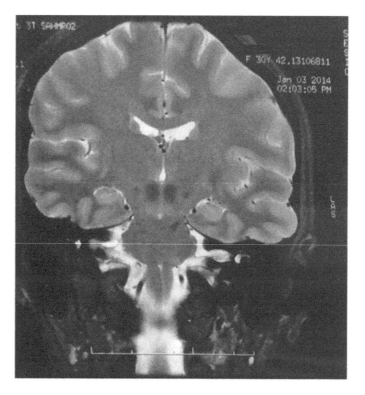

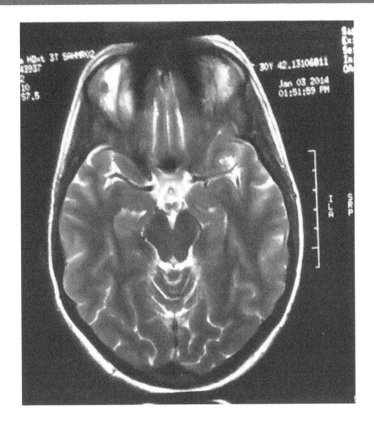

A.3 Human Ear

Command terms: draw, label and annotate

Label the diagram of the human ear.

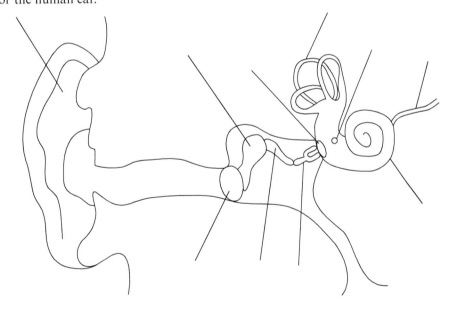

A.3 Perception of Auditory Stimuli

Command terms: label, annotate and outline

Place the following parts of the human ear into the correct order in the flow chart to represent the passage, amplification and transmission of sound waves.

Indicate with ★ where amplification of sound occurs.

Structures
Cochlea, stapes, tympanic membrane, oval window, auditory nerve, incus, pinna, malleus, cilia of hair cells

A.3 Types of Receptors

Command terms: define, state and identify

Complete the table below to give at least two examples for each type of receptor.

Type of receptor	Description	Examples
Chemoreceptor	Odours Taste	
Thermoreceptor	Heat Cold	
Mechanoreceptor	Motion Sound Touch Pressure Stretching	
Photoreceptor	Light	

A.3 Human Eye

Command terms: draw, label and annotate

Label the diagram of the human eye.

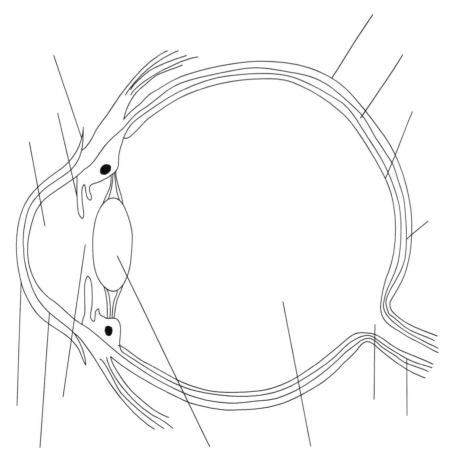

A.3 The Retina

Command terms: draw, label and annotate

Label and annotate the diagram of the retina to show the cell types.

Draw an arrow to show the direction in which light moves.

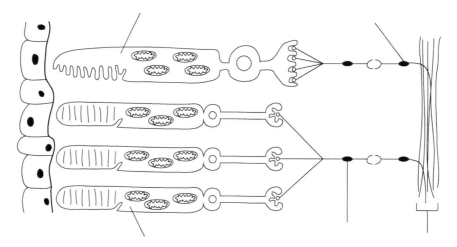

A.4 Reflex Arc

Command terms: draw, label, state, annotate and identify

Complete the diagram below to show the path of a pain withdrawal reflex arc through the spinal cord. Draw and label the parts listed in the box.

Receptor cell
Sensory neuron
Relay neuron
Motor neuron
Effector
White matter
Grey matter
Ventral root
Dorsal root

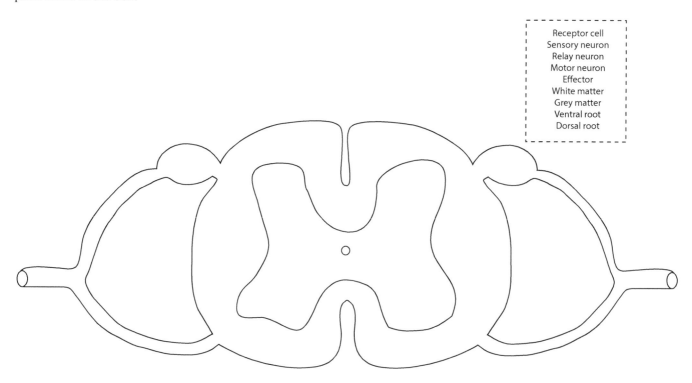

A.4 Innate and Learned Behaviour

Command terms: define, state and identify

Connect the following terms with their description.

Innate behaviour	The process of encoding, storing and accessing information
Learned behaviour	An involuntary response
Autonomic response	Behaviour that is inherited from parents and therefore develops independently from environmental stimuli
Imprinting	A rapid, unconscious and involuntary response to an environmental stimulus
Operant conditioning	Learning that occurs at a particular life stage and is independent of the consequences of the behaviour
Reflex	Behaviour that develops as a result of experience
Memory	A form of learning that consists of trial and error experiences

A.4 Pavlov's Experiments

Command terms: state, describe and outline

Fill in the blanks to complete the sentences below that outline Pavlov's experiments into reflex conditioning in dogs.

Ivan Pavlov investigated the _____ in dogs.

The dogs were presented with _____ at the same time as a _____ was rung.

The dogs learnt to associate the sound of the _____ with the _____ food.

After several repeats, the dogs _____ after only the _____ _____ even in the absence of _____.

The smell or sight of the food was the _____.

The salivation in response to this food was the _____.

The sound of the bell was the _____.

The salivation at the sound of the bell was the _____.

This is an example of _____.

Pavlov also found that other signals could be used instead of the bell, including _____, _____ _____ and _____.

A.5 Stimulants and Sedatives

Command terms: define, identify, compare and explain

Complete the following questions relating to two different classes of drugs.

1. Drugs can be classed as stimulants or sedatives. Define these terms.

    Stimulant

    _____

    _____

Sedative

_____

_____

2. Identify the following drugs as being either a stimulant or a sedative.

Nicotine   _____

Alcohol   _____

Tetrahydrocannabinol (THC)   _____

Cocaine   _____

Benzodiazepines   _____

Amphetamines   _____

3. Complete the following table to explain the effects of any two stimulants and any two sedatives on the nervous system.

Stimulant	Effect on the nervous system

Sedative	Effect on the nervous system

A.6 Ethology

Command terms: describe, outline and discuss

Complete the table below to summarise the examples of animal behaviour listed. Provide examples wherever possible.

Behaviour type	Species	Description
Migratory	Blackcaps	
Blood sharing	Vampire bats	
Foraging	Shore crabs	
Breeding strategies	Coho salmon	
Courtship	Birds of paradise	
Synchronised oestrus	Female lions	
Feeding	Blue tits	

BIOTECHNOLOGY AND BIOINFORMATICS (OPTION B)

B.1 Fermentation

Command terms: list, state and outline

Complete the table below to outline the optimum conditions required in industrial fermenters and the factors limiting fermentation.

Optimum conditions required in fermenters	Factors limiting fermentation in fermenters

B.1 Gram Staining

Command terms: distinguish and compare

Gram staining is a test used to classify bacteria into either Gram-negative or Gram-positive based on how they react to the stain. Complete the following table to compare Gram-negative bacteria with Gram-positive bacteria.

	Gram negative	Gram positive
Gram reaction		
Cell wall thickness		
Thickness of peptidoglycan layer		
Outer membrane present?		
Acidic polysaccharides present?		
Lipopolysaccharides present?		

B.1 Biogas Production

Command terms: describe, outline and explain

Bacteria and archaeans produce biogas from organic matter in fermenters. Number the steps below in the correct order to show the process of biogas formation in fermenters.

Slurry is removed from the fermenter and used as fertiliser _____

Methanogens produce methane by reducing carbon dioxide to methane _____

Raw organic waste is converted into organic acids, alcohol, hydrogen and carbon dioxide _____

Methane leaves the fermenter for use in heating or cooking _____

Methanogens produce methane by splitting ethanoic acid to form carbon dioxide and methane _____

Organic acids and alcohol are used to produce acetate, hydrogen and carbon dioxide _____

Sewage, manure and other organic wastes are fed into the fermenter _____

B.2 Recombinant DNA

Command terms: list, state and identify

Determine which of the following statements about recombinant DNA are true or false.

	True/False
Recombinant DNA contains genetic material from one or more sources	_____
Recombinant DNA is only produced using genetic material from bacteria or viruses	_____
Recombinant DNA can only be inserted into plant cells	_____
A host cell is required for recombinant DNA to be inserted	_____
Chromosomes are essential for the expression of the new gene	_____
Plant cells can have DNA inserted into chloroplasts	_____
A vector is not essential in transformation	_____
Gene flow between genetically modified plants and non-genetically modified plants is important	_____
Recombinant DNA can be introduced into whole plants, leaf discs or protoplasts	_____
Recombinant DNA can be used to overcome problems of environmental resistance	_____
Recombinant DNA can cause problems of environmental resistance in crops	_____
Recombinant DNA is only introduced chemically into plant cells	_____

B.2 Examples of Genetic Modification

Command terms: outline, discuss and explain

The table below gives examples of genetic modification currently in use today. Complete the table by outlining the benefit and risk associated with each example.

Example	Benefit	Risk
Introduction of glyphosate resistance in soybean crops		
Production of Hepatitis B vaccine in tobacco plants		
Production of Amflora potato for paper and adhesive industries		

B.3 Examples of Bioremediation

Command terms: outline, discuss and explain

Bioremediation strategies involve the use of microorganisms to clean up or control environmental contaminants. The table below gives examples of bioremediation strategies currently in use today. Complete the table by explaining the benefits associated with each example.

Example	Benefit
Degradation of benzene by halophilic bacteria	
Degradation of oil by *Pseudomonas*	
Conversion of methyl mercury into elemental mercury by *Pseudomonas*	

B.3 Biofilms

Command terms: state, describe, outline and discuss

Microorganisms can form colonies called biofilms, which give them different properties as a whole colony than those of each individual organism. Complete the following table to summarise the advantages and disadvantages of biofilms.

Advantages of biofilms	Disadvantages of biofilms

B.4 Interpretation of ELISA

Command terms: measure, calculate, estimate, determine and predict

ELISA = enzyme-linked immunosorbent assay

The purpose of ELISA is to determine if a protein is present and how much of that protein is present in a sample. It is used to test for the presence of infection caused by a pathogen.

Antibodies are used and a colour change identifies the substance of interest. A positive test is indicated by this colour change, whilst in a negative test, the samples remain uncoloured.

Examples of application include:
- Testing for HIV, Lyme disease, malaria, Chagas disease
- Detection of rotavirus, hepatitis B, squamous cell carcinoma, syphilis, coeliac disease
- Detecting potential food allergens (e.g. milk, nuts and eggs)
- Toxicology testing for certain drugs (e.g. benzodiazepines, cannabinoids)

B.4 Gene Therapy

Command terms: list and state

Gene therapy can be used to treat or even cure some genetic diseases by replacing defective copies of genes with properly functioning ones. There are two types of gene therapy, somatic therapy and germ line therapy.

Complete the following table to list genetic conditions that may be treated or cured using gene therapy.

Examples of genetic conditions treated with gene therapy

B.5 Databases

Command term: analyse

The following list of databases can be used by scientists or the general public to access information and analyse sequence data in biological research.

Database	Web address	Use
BLAST	http://blast.ncbi.nlm.nih.gov/Blast.cgi	Nucleotide sequences, protein sequences, translated nucleotides
GenBank	http://www.ncbi.nlm.nih.gov/genbank/	Genetic sequence search
DDJB	http://www.ddbj.nig.ac.jp	Nucleotide sequences
EMBL (ENA)	http://www.ebi.ac.uk/ena	Nucleotide sequences
SwissProt	http://www.ebi.ac.uk/uniprot	Protein sequences
PIR International	http://pir.georgetown.edu	Protein sequences
NCBI	http://www.ncbi.nlm.nih.gov	BLAST searches and construction of phylograms and cladograms
ArrayExpress	https://www.ebi.ac.uk/arrayexpress/	Gene expression data from microarray studies
KEGG	http://www.genome.jp/kegg/	Genome maps, gene sequences, human diseases
PDB	http://www.wwpdb.org	Information about protein structure and nucleic acids
Ensembl	www.ensembl.org/	Gene sequences, coding and non-coding sequences on chromosomes

CHAPTER 14 ECOLOGY AND CONSERVATION (OPTION C)

C.1 Species and Communities

Command terms: define and state

Link the following terms with their meanings.

Limiting factor	The position and distribution of a population within their habitat.
Keystone species	The actual part of the niche that an organism occupies taking into account the presence of any limiting factors.
Spatial habitat	A relationship of mutual benefit.
Fundamental niche	Factors such as the availability of food, water and shelter limit the growth of a species within an ecosystem.
Realised niche	A species with a disproportionally large impact on its surrounding environment in relation to the abundance of the species.
Symbiotic relationship	The full range of resources an organism can use and environment it can occupy within its habitat, in the absence of limiting factors.

C.2 Climograph Analysis

Command terms: state, identify, analyse, deduce and determine

A climograph shows the mean precipitation and mean temperature conditions needed in a particular ecosystem. Each ecosystem type has a different combination of climatic conditions. You are required to analyse a climograph and answer questions such as those below. For an example climograph visit http://www.zo.utexas.edu/courses/bio301/chapters/Chapter4/Chapter4.html

1. Determine the range of mean temperature and precipitation conditions required in a temperate forest ecosystem.

2. State all types of ecosystem that can form when mean annual temperatures are **20°C**.

3. State all types of ecosystem that can form when mean annual precipitation levels are **100 cm**.

_____

_____

_____

4. Compare the conditions required for a tropical rainforest to form with those required for a temperate rainforest.

_____

_____

_____

_____

_____

5. Some climographs have the edges of ecosystem types shown with dotted lines. Explain the reason for these dotted lines.

_____

_____

_____

C.2 Food Webs

Command terms: state and identify

The box below has information about species found in an Australian grassland ecosystem.

Plants:

Golden Wattle tree (*Acacia pycnantha*)
Lemon-scented Eucalyptus tree (*Eucalyptus citriodora*)
Kangaroo grass (*Themeda australis*) and
Golden Beard grass (*Chrysopogon fallax*)

Animals:

Blue-faced honeyeaters (*Entomyzon cyanotis*) feed on nectar and sap from *E. citriodora*
Termites (*Mastotermes darwiniensis*) feed on *E. citriodora*, *A. pycnantha* and *T. australis*
Field crickets (*Lepidogryllus comparatus*) feed on *T. australis*
Hairy-nosed wombats (*Lasiorhinus krefftii*) feed on *C. fallax*
Eastern Grey kangaroos (*Macropus giganteus*) feed on *T. australis*
Emus (*Dromaius novaehollandiae*) feed on *A. pycnantha* and *L. comparatus*
Short-beaked echidnas (*Tachyglossus aculeatus*) feed on *M. darwiniensis*
Magpies (*Cracticus tibicen*) feed on *M. darwiniensis* and *L. comparatus*
Frilled-neck lizards (*Chlamydosaurus kingii*) feed on *M. darwiniensis* and *L. comparatus*
Dingoes (*Canis lupus dingo*) feed on *M. giganteus*, *D. novaehollandiae*, *L. krefftii* and *C. kingii*
Wedge-tailed eagles (*Aquila audax*) feed on *M. giganteus* and *C. kingii*
Laughing kookaburras (*Dacelo novaeguineae*) feed on *C. kingii*

Use this information to create a food web connecting each of the species listed above.

C.2 Gersmehl Diagrams

Command terms: annotate, identify, outline, construct and deduce

The following Gersmehl diagrams depict the nutrient storage and nutrient flow for three different terrestrial ecosystems.
S = soil   L = litter   B = biomass.

1. Identify which ecosystem each diagram represents; taiga, desert or tropical rainforest.
2. Annotate each diagram by adding detail to the arrows demonstrating nutrient flow.

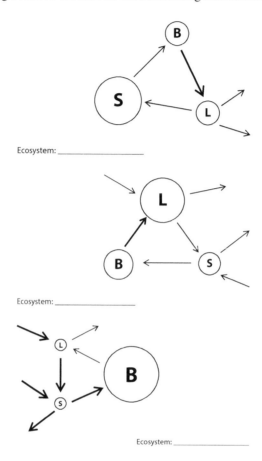

Ecosystem: _____

Ecosystem: _____

Ecosystem: _____

C.3 Biomagnification

Command terms: list, state and compare

Biomagnification is the increase in concentration of chemical substances as you move up the food chain from one trophic level to the next. Complete the table to summarise the causes and consequences of biomagnification on individuals, species and ecosystems. Give specific examples where possible.

Causes of biomagnification	Consequences of biomagnification

C.4 Simpson's Diversity Index

Command terms: measure, calculate, estimate, determine and predict

Simpson's reciprocal index of diversity is a measure of diversity in ecology, which is used to quantify the biodiversity of a habitat.

The formula for Simpson's diversity index is:

$$D = \frac{N(N-1)}{\Sigma n(n-1)}$$

n = number of individuals of a particular species

N = total number of organisms of all species found

The greater the biodiversity in an area, the higher the value of D. The lowest possible value for D occurs when only one species is present within the area being studied (value of $D = 1$).

The table below shows data collected at two sites in eucalypt forest in the Mount Lofty ranges of South Australia. A total of 50 butterflies were counted at each site, with the number of individuals from each species shown in the table.

Species of butterfly	Number of individuals at site X	Number of individuals at site Y
Heteronympha merope merope	16	20
Polyura sempronius	13	21
Vanessa itea	5	1
Danaus plexippus plexippus	12	2
Zizina labradus labradus	4	6

1. Calculate Simpson's reciprocal diversity index (D) for the butterflies found at each site.

_____

_____

_____

2. Compare the butterfly populations at site X with site Y.

_____

_____

_____

3. Suggest a possible conclusion that could be drawn from this study.

_____

_____

_____

_____

C.4 In-Situ versus Ex-Situ Conservation

Command terms: define, list, state, compare

Complete the table below to compare in-situ conservation and ex-situ conservation and provide examples for each type.

	In-situ conservation	Ex-situ conservation
Definition		
Benefits		
Disadvantages		
Examples		

C.5 Population Growth Curves

Command terms: measure, calculate, estimate, determine and predict

The graph below shows the population growth of the Dusky Hopping Mouse (*Notomys fuscus*) in the Strzelecki Desert in central Australia over the past 60 years.

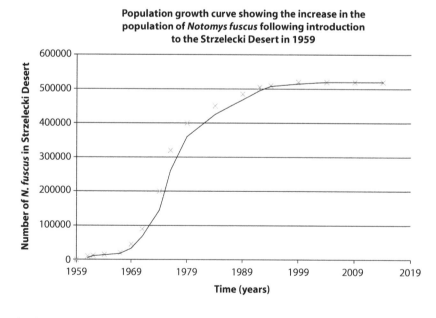

Population growth curve showing the increase in the population of *Notomys fuscus* following introduction to the Strzelecki Desert in 1959

1. Indicate on the graph where the following phases occur:

    Initial phase
    Exponential growth
    Transitional phase
    Plateau/carrying capacity

2. Outline how each of the phases shown can be explained by relative rates of natality, mortality, immigration and emigration.

_____

_____

_____

_____

3. Discuss the factors that may have set limits to the population increase of *N. fuscus*.

_____

_____

_____

_____

C.5 The Lincoln Index

Command terms: calculate and estimate

One method used to sample animal populations and determine population size is the capture–mark–release–recapture method. Once field data have been collected in this way, the Lincoln index can be applied to estimate the size of the population being studied.

1. As many as possible individuals of the target species are captured.
2. All captured individuals are marked in a manner that leaves the individual unharmed and does not attract predators.
3. The marked individuals are released and allowed sufficient time to settle back in to their habitat.
4. Later, as many individuals as possible are captured from the same place as the first capture.
5. The numbers of marked and unmarked individuals captured in the second batch are counted.
6. The Lincoln index is applied and the population size can be estimated.

The formula for the Lincoln index is:

$$\text{Population size} = \frac{n_1 \times n_2}{n_3}$$

n_1 = number initially caught and marked

n_2 = total number caught on second occasion

n_3 = number of marked individuals recaptured

C.6 The Nitrogen Cycle

Command terms: draw, label, state, annotate, identify and outline

Use arrows to link the boxes and label these arrows using the descriptions given to the right.

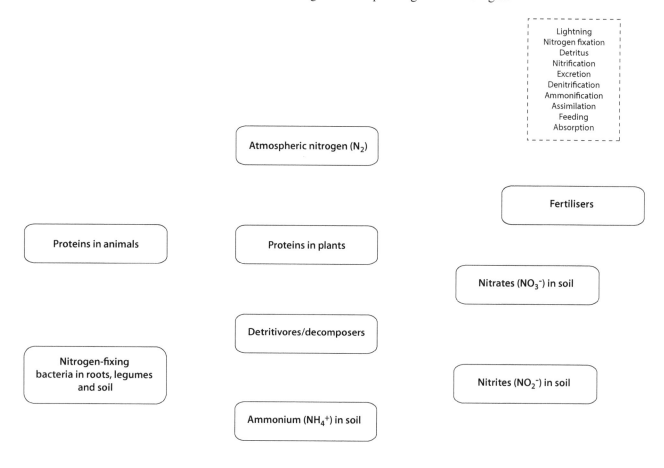

C.6 The Phosphorus Cycle

Command terms: state, describe, outline and explain

Fill in the blanks to complete the sentences below that outline the phosphorus cycle.

Phosphorus is used to make molecules including _____, _____ and _____.

It also makes up the _____ of the cell's plasma membrane.

Phosphorus occurs in many different forms.

Phosphorus can be _____ in waterways and oceans after running off the land.

Phosphorus is present in soil as _____ after _____ break down animal and plant wastes.

Rocks contain phosphorus in the form of _____ after geologic uplift from deep ocean _____ _____.

Phosphates are readily taken up by _____ to enter the food chain.

Phosphorus has a low _____ when compared with nitrogen.

Human activity affects the phosphorus cycle through the addition of phosphate _____ or removal by harvesting of _____.

CHAPTER

15 HUMAN PHYSIOLOGY (OPTION D)

D.1 Nutrients Synthesised versus Dietary Requirement

Command terms: list, state, distinguish, identify and compare

The table below summarises those nutrients made by the human body and those not synthesised by the body and therefore included in the diet.

Synthesised by the body	Dietary requirement
Omega-9 fatty acids Carbohydrates such as glucose Vitamins B7 (biotin), D (calciferol) Protein Cholesterol Amino acids such as aspargine, aspartic acid, I-cysteine, glycine, serine, tyrosine, glutamic acid, proline, glutamine and alanine	**Vitamins** A (retinol), B1 (thiamine), B2 (riboflavin), B3 (niacin), B5 (panthothenic acid), B6 (pyridoxine), B9 (folic acid), B12 (cobalamin), C (ascorbic acid), E (tocopherol), K (naphthoquinoids) **Essential fatty acids** α-Linolenic acid (an omega-3 fatty acid), linoleic acid (an omega-6 fatty acid) and arachidonic acid **Essential amino acids** Isoleucine, lysine, leucine, methionine, phenylalanine, threonine, tryptophan, valine, histidine and arginine **Minerals** Calcium, chloride, chromium, cobalt, copper, iodine, iron, magnesium, manganese, molybdenum, phosphorus, potassium, selenium, sodium, zinc **Other** Choline

D.2 Hormonal and Nervous Control of Digestion

Command terms: distinguish, outline and compare

Complete the table below by correctly assigning each stage in the secretion of digestive juices as either being under nervous control or hormonal control.

Nervous control	Hormonal control

Sustained release of gastric juice

Brain is stimulated to send nerve impulses to exocrine glands in the stomach wall

Gastrin stimulates exocrine glands to secrete hydrochloric acid

Release is as a result of touch receptors, chemoreceptors and stretch receptors detecting the food when it reaches the stomach

Brain sends more nerve impulses to exocrine glands

Secreted before food even reaches the stomach

Endocrine glands secrete the hormone gastrin

Initial release of gastric juice

A reflex action to the sight or smell of food

D.2 Structure and Function of Villus Epithelial Cells

Command terms: state, describe and outline

Complete the table below to outline the function of the named structures in the absorption of food in the intestinal villi.

Structure	Function
Microvilli	
Mitochondria	
Pinocytic vesicles	
Tight junctions	

D.3 Breakdown of Haemoglobin

Command terms: label, annotate, outline and construct

Complete the flow chart to outline the process of erythrocyte and haemoglobin breakdown in the liver.

Annotate the arrows to describe the process occurring at each stage or the location or cells involved.

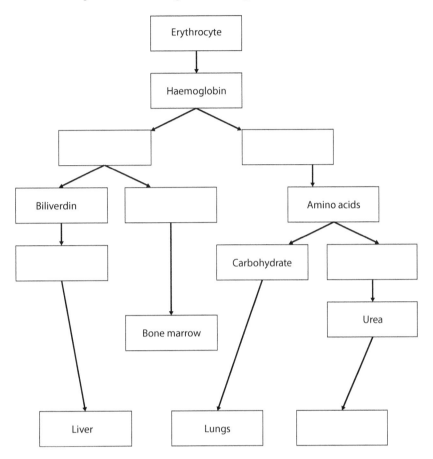

D.3 Jaundice

Command terms: state and list

The table below summarises the causes and consequences of jaundice in infants and adults.

Causes of jaundice	Consequences of jaundice
Increased levels of bilirubin in the blood (hyperbilirubinemia)	Kernicterus (in infants) resulting in brain damage, deafness and cerebral palsy
Hepatitis	Itchiness (in adults)
Liver cirrhosis	Yellowing of skin and eyes
Alcoholic liver disease	
Drug-induced hepatitis	
Liver cancer	
Leptospirosis	
Obstruction of bile duct (e.g. from gallstones or pancreatic cancer)	
May be congenital	
Common in newborns due to high turnover of red blood cells, bilirubin not processed quickly enough or lack of absorption of bilirubin	

D.3 Comparison of Capillaries and Sinusoids

Command terms: distinguish, outline and compare

The table below summarises the similarities and differences in normal capillaries and the sinusoids of the liver.

Capillaries	Sinusoids
Composed of endothelium	Composed of endothelium
Narrow openings	Wider openings
Walls consist of single layer of very thin cells	Walls consist of single layer of very thin cells
Have a basement membrane	No basement membrane
Endothelial cells overlap so less or no gaps between endothelial cells	Many open pores between endothelial cells (fenestrated)
Many tight junctions	No or few tight junctions
Have pinocytotic vesicles	Lack pinocytotic vesicles
	Walls are more porous

D.3 Blood Supply to the Liver

Command terms: label, annotate and identify

1. Label the blood vessels.
2. Annotate the diagram to outline the origin of blood flowing through each of the blood vessels.
3. Colour the blood vessels to show whether they carry oxygenated (red) blood or deoxygenated (blue) blood.

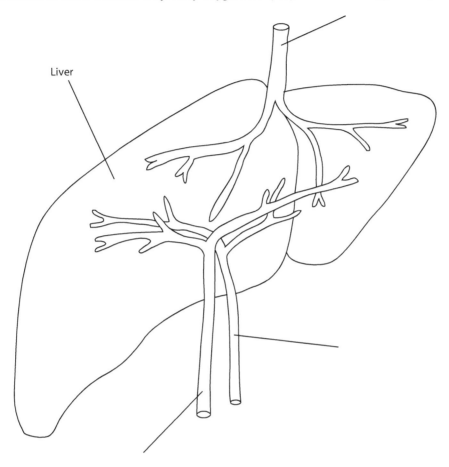

Liver

D.4 Cardiac Cycle

Command terms: outline, describe and explain

Place the events of the cardiac cycle in the correct order by numbering the events.

Blood enters the atria	_____
Atria and ventricles are relaxed	_____
SA node initiates impulse	_____
Atria contract	_____
Blood is forced into the ventricles	_____
Atrioventricular (AV) valves close	_____
Ventricles contract	_____
Blood flows to aorta and pulmonary arteries	_____
Semilunar valves close	_____
Blood returns to left atrium from pulmonary vein	_____

D.4 Mapping the Cardiac Cycle

Command terms: draw, label, annotate, identify, deduce and sketch

The following graph shows the results of a normal electrocardiogram (ECG).

Map the cardiac cycle on the graph by adding annotations and labels.

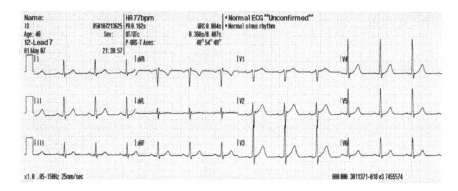

Source: Glenlarson, 2007, *Lead Generated Sinus Rhythm*, http://commons.wikimedia.org/wiki/File%3A12_lead_generated_sinus_rhythm.JPG.

D.5 Steroid versus Peptide Hormones

Command terms: state, distinguish, outline and compare

Complete the table below by assigning the example hormones given below to the correct category, either steroid hormone or peptide hormone.

Type of hormone	Mode of action	Examples
Steroid hormone	Enter target cells through the plasma membrane Bind to receptors in the cytoplasm Interact directly with genes Control activity and development of target cells Require carrier proteins to travel in the blood	
Peptide hormone	Bind to receptors within the plasma membrane Cause release of secondary messenger inside the cell Causes changes in or inhibits enzyme activity Do not enter the cell	

Oestrogen
Oxytocin
Follicle-stimulating hormone (FSH)
Anti-diuretic hormone (ADH)
Luteinising hormone (LH)
Glucagon
Progesterone
Cortisol
Thyroid-stimulating hormone (TSH)
Testosterone
Insulin
Gastrin
Growth hormone (GH)
Leptin
Aldosterone
Human chorionic gonadotropin

D.5 Pituitary Hormones

Command terms: state and list

The hormones listed below are produced either by the anterior pituitary or posterior pituitary. Correctly assign them to the correct category.

Anterior pituitary	Posterior pituitary

Prolactin
Oxytocin
FSH
LH
TSH
ADH
Adrenocorticotropic hormone (ACTH)
Melanocyte-stimulating hormone (MSH)
GH

D.6 Oxygen Dissociation Curves

Command terms: label, annotate, construct and sketch

Complete the graph to show the oxygen dissociation curves for adult haemoglobin, myoglobin and foetal haemoglobin. Use a different colour for each.

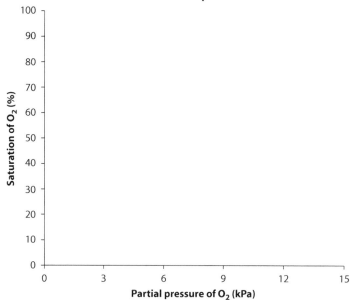

Oxygen dissociation curves for adult haemoglobin and myoglobin and foetal haemoglobin in *Homo sapiens*

Appendix 1
IMPORTANT BIOLOGICAL PREFIXES AND SUFFIXES

Prefixes

A

Ab-	away from
A-/An-	without, negative
Ad-	toward, near
Aden-	gland
Adren-	towards the kidney
Aero-	air, with oxygen
Agon-	contest
Allo-	different
Alveol-	cavity
Ama-	together
Amyl-	starch
Ana-	up
Andr-	man
Angi-	vessel
Ant-	against, opposing
Ante-	before
Anth-	flower
Api-	tip, extremity
Apo-	off, away
Arthr-	joint
Auto-	self
Aux-	growth, increase

B

Basal-	base
Bi-	two
Bio-	life
Blast-	bud, germ
Blasto-	produce
Brachi-	arm
Broncho-	bronchus
Bryo-	moss

C

Calor-	heat
Capill-	hair
Capsa-	a box
Carcin-	cancer
Cardio-	heart
Carp	fruit
Cata-	down, lower
Centro-	centre
Cephal-	head
Cerebro-	brain
Chem-	chemical
Chiasm-	crossing
Chlor-	green
Chrom-	colour
Cili-	small hair
Circ-	circle
Clado-	branch
Co-	with, together
Coch-	a snail
Coll-	glue
Com-	together
Contra-	against, preventing
Counter-	opposite
Corp-	body
Cost-	rib
Crani-	skull
Cusp-	pointed
Cut-	skin, integument
Cycl-	circular
Cyst-	sac, bladder
Cyt-/Cyto-	cell

D

Dactyl-	finger, toe, digit
De-	from, opposite
Decid-	falling off
Demi-	half
Dendr-	tree, branch
Dent-	tooth
Deoxy-	have less oxygen
Derm-	skin
Detrit-	wear off
Di-/Dipl-	double, twice, two

Dia-	through, between
Dis-	negative, opposite
Dorm-	sleep
Dors-	the back
Dys-	abnormal, faulty

E

Eco-	environment
Ecto-	outer, external
Ef-	out, away
Em-/En-	into, inside, within
Embry-	egg
Enceph-	brain
End-/Endo-	inside, within
Entero-	intestines, gut
Ep-/Epi-	upon, over, above
Equi-	equal, alike
Erythro-	red
Etho-	custom, habit
Etio-	cause
Eu-	good, well, true
Ex-/Exo-	out of, external
Excit-	stimulating
Extra-	outside, beyond

F

Fenestr-	window
Ferr-	iron
Fertil-	fruitful
Fibro-	fiber
Fil-	thread
Flagell-	whip
Folli-	bag
Fore-	before, front

G

Galact-	milk
Gam-	united, joined
Gangli-	swelling, knot
Gastr-	stomach
Gen-	origin, produce
Geno-	offspring
Geo-	the Earth
Germin-	grow
Gest-	carried
Glom-	ball
Gluco-	sweet
Glyco-	sugar, sweet
Gon-/Gono-	seed, offspring
Gust-	taste
Gyn-	women

H

Haem-/Hem-	blood
Halo-	salt
Hapl-	single
Helic-	a spiral
Hemi-	half, partial
Hepat-	liver
Hetero-	other, different
Hist-	tissue
Holo-	whole
Homeo-	same
Homo-	same, similar
Hydra-/Hydro-	water
Hyper-	over, above, excess
Hypo-	under, deficient
Hyster-	uterus

I

Im-/In-	in, into, inward, not
Infra-	below, inferior
Inter-	between, among
Intero-	inside
Intra-	among, within
Intro-	within
Immuno-	immune system
Iso-	equal

K

Karyo-	nucleus
Kerat-	horn
Kin-	movement, action

L

Lact-	milk
Lamin-	sheet, layer
Later-	side
Leuk-/Leuc-	white
Liga-	bound, tied
Lip-/Lipo-	fat, fatty
Lumen-	light
Lute-	yellow
Lymph-	lymphatic system
Lyso-/Lyto-	loosen

M

Macro-	large
Mal-	bad, abnormal
Mamm-	breast, teat

Mater-	mother
Medi-	middle
Meio-	less
Melan-	black, dark
Mening-	membrane
Ment-	mind
Meso-	middle
Meta-	beyond, between
Micro-	small
Mono-	single, one
Morph-	form, shape
Muc-/Muco-	mucous
Multi-	many
Muta-	change
Myelo-	spinal cord
Myo-	muscle

N

Nas-/Naso-	nose
Necro-	death
Neo-	new
Neph-	kidney
Neuro-	nerve
Nitr-	nitrogen, nitrate
Non-	not
Noto-	back
Nucle-	nucleus
Nutri-	feed, nourish

O

Ob-	before, against
Oculo-	eye
Olfact-	smell
Omni-	all
Onc-	tumour, cancer
Oo-/Ovi-	egg, ovum
Ophthalm-	eye
Org-	living
Ortho-	straight, normal
Osmo-	osmosis, pushing
Ost-/Osteo-	bone
Oxy-	oxygen

P

Paleo-	ancient
Para-	beside, near
Path-	disease
Pep-	digest
Per-	through
Peri-	around
Phag-	feeding, ingesting
Pheno-	show, appear

Phil-	love
Phleb-	vein
Phob-	fear
Phon-	sound
Photo-	light
Phren-	mind
Phyll-	leaf
Phyt-	plant
Pil-	hair
Pin-/Pino-	drink
Platy-	flat
Pleur-	side, rib
Pluri-	more, several
Pneu-	air, lungs, wind
Pod-	foot
Poly-	many, multiple
Post-	after, behind
Pre-	in front of, ahead
Pro-	before, earlier than
Proto-	first
Pseud-	false
Psych-	the mind
Pulmo-	lung

Q

Quadr-	four

R

Radia-	spoke, ray
Re-	back, again
Ren-	kidney
Retic-/Retin-	network
Retro-	backward, behind
Rhiz-	root

S

Sapro-	decay, rotten
Sarco-	flesh, soft tissue
Sclero-	hard
Semi-	half
Sin-/Sino-	a hollow
Soma-	body, of the body
Sperm-	seed, sperm
Spiro-	respiration
Spor-	seed, spore
Stoma-	mouth, opening
Sub-	beneath, under
Sucr-	sugar, sweet
Super-	above, upon
Syn-/Sym-	with, together
Systol-	contraction

T

Tachy-	rapid
Tact-	touch
Tax-	arrangement
Tel-	end, completion
Tens-	stretched
Tetra-	four
Therm-	heat, hot
Thorac-	chest
Thromb-	clot
Thyr-	thyroid gland
Tono-	stretched
Tox-	poisonous, toxic
Trans-	across
Tri-	three, triple
Troph-	nourish
Tuber-	swelling
Turg-	swollen
Tympan-	a drum

U

Ultra-	beyond, in excess
Un-	not
Uni-	one, single
Ur-	urine

V

Vacu-	empty
Vagin-	sheath
Vasa-	vessel
Ven-	vein
Viri-	virus
Viscero-	internal organs
Vit-	life
Vitre-	glass, glassy
Viv-	alive, living
Vora-	eat

X

Xer-	dry
Xyl-	wood

Z

Zo-/Zoo-	animals
Zona-	a belt
Zyg-	union, fusion
Zym-	ferment, enzyme

Suffixes

A

-able	capable of
-ac	affected by, refer to
-al	of, belonging to
-angio	vessel
-aphy	suck
-apsis	juncture
-ary	associated with
-ase	enzyme

B

-bios	life
-blast	bud, sprout
-bryo	swollen

C

-centesis	a puncture
-cide	destroy, kill
-clin	slope

-crine	separate
-cutane	skin
-cycle	circle
-cyte	cell

D

-derm	skin
-duct	to lead

E

-ectomy	cutting out
-ell/-elle	small
-emia	blood

F

-ferent	carry, bring
-form/-forma	shape
-fuge	driving out

G

-gamy	reproduction
-gen	initiating agent
-genesis	origin, birth
-genic	producing
-geny	origin
-glia	glued together

I

-ia	condition
-ism	condition
-itis	inflammation

K

-kinesis	motion, movement

L

-lemma	sheath
-lite	first
-logy	the study of
-lyse	break
-lysis	split, break apart

M

-mere	part

N

-nata	birth

O

-oid	like, resembling
-ology	the study of
-oma	tumor
-opia	defect of the eye
-ory	referring to
-ose	sugar
-osis	affected, abnormal
-oxide	containing oxygen

P

-pathy	disease
-patri	father
-phage	to eat
-phil	like, love
-phyll	keaf
-phragm	partition
-plas	grow
-plast	formed, molded
-pnea	air, breathing
-port	gate, door
-pter	wing, feather, fin

R

-rrhea	flow, discharge

S

-sacchar	sugar
-some	body
-sorb	suck in
-stalsis	constriction
-stasis	arrest, fixation

T

-tonus	tnesion
-tomy	to cut
-topo	place
-tropic	turn, change
-ty	condition of, state

U

-uria	urine

V

-valent	strength
-vect	carried

Y

-yl	substance, matter

Appendix 2
DEFINITIONS OF KEY BIOLOGICAL TERMS

Command term: define

A

Abiotic (SL)	The non-living components of an ecosystem such as climate and the availability of resources such as water and food.
Absorption (SL)	The taking in of chemical substances through cell membranes or layers of cells.
Absorption spectrum (SL)	A graph showing the degree of absorbance of different wavelengths of light by chlorophyll during photosynthesis.
Actin (HL)	A protein that forms microfilaments and is a large component of the cytoskeleton and involved in skeletal muscle contraction.
Action potential (SL)	The localised reversal and then restoration of electrical potential between the inside and outside of a neuron as the impulse passes along it.
Action spectrum (SL)	A graph showing the efficiency of photosynthesis at different wavelengths of light.
Activation energy (HL)	The required level of energy needed for a chemical reaction to take place.
Active immunity (HL)	Immunity due to the production of antibodies by the organism itself after the body's defence mechanisms have been stimulated by antigens.
Active site (SL)	A region on the surface of an enzyme to which substrates (reacting substance) bind and which catalyses a chemical reaction involving the substrates.
Adenosine tri-phosphate (ATP) (SL)	A molecule that transports chemical energy within cells and is involved in cell metabolism.
Adhesion (SL)	The tendency of molecules to be attracted to other molecules.
Aerobic (SL)	In the presence of oxygen, as in aerobic cellular respiration.
Affinity (HL-D)	The attraction of one molecule to another.
Alien species (SL-C)	A species that is introduced by humans to an area in which it does not naturally occur either by accident or on purpose.
Allele (SL)	One specific form of a gene, differing from other alleles by one or a few bases only and occupying the same gene locus as other alleles of the gene.
Amniocentesis (SL)	A procedure whereby a small amount of amniotic fluid is collected from the amniotic sac of a pregnant woman and then tested to diagnose chromosomal abnormalities in the foetus.
Anabolism (SL)	The building of large molecules from smaller molecules.
Anaerobic (SL)	Without the presence of oxygen, as in anaerobic cellular respiration.
Analogous trait/structure (SL)	A trait or structure that performs the same or similar function but has different evolutionary origins.
Antagonistic (HL)	Muscles occurring pairs, with each muscle of the pair having an opposite effect to the other muscle in the pair.
Anterior (SL-A)	Relating to the front of the individual.
Antibodies (SL)	Proteins produced in response to an antigen to neutralise the effect of that antigen (antibodies recognise and bind to specific antigens).
Anticodon (SL)	The triplet of bases found on the tRNA that corresponds to the complementary codon on the mRNA.
Antigens (SL)	Any foreign particle that causes the body to produce antibodies.

Antiparallel (SL)	The DNA molecule has two strands that run in opposite directions to one another, one being in the 3'–5' direction and the other being in the 5'–3' direction.
Antisense strand (HL)	The template strand that is transcribed.
Artificial selection (SL)	The selection by humans of desired traits in an organism and breeding those organisms with such traits so that these traits are seen in the offspring.
Assimilation (SL)	The conversion of nutrients into protoplasm that in animals follows digestion and absorption.
Autonomic (SL-A)	Part of the peripheral nervous system that is in control of involuntary or unconscious processes.
Autosomes/autosomal (SL)	All the chromosomes that are not sex chromosomes.
Autotroph (SL)	An organism that synthesizes its organic molecules from simple inorganic substances.
Axon (SL-A)	The long, thin part of a neuron that conducts electrical impulses during an action potential.

B

Base substitution mutation (SL)	A mutation occurring when one base is substituted for a different base.
Binary fission (SL)	The process of cell division undertaken by prokaryotes that allows doubling of the number of cells with each split.
Biofilm (SL-B)	A group of microorganisms that stick to one another and then to a surface.
Biomagnification (SL-C)	The presence of a chemical that, when taken into a food chain, accumulates in the tissue of the organism that has ingested it and is passed on to those organisms further up the food chain. Thus those at the top of the food chain are most affected.
Biomass (SL)	The material derived from living organisms or recently living organisms.
Biopharming (HL-B)	The use of genetically modified crops as drug-producing bioreactors to produce vaccines and medicines.
Bioremediation (SL-B)	The use of organisms such as bacteria and fungi to neutralise or remove contaminants or pollutants from an affected site.
Biotic (SL)	The living component of an ecosystem such as the presence of mates, predators or prey.
Bohr shift (HL-D)	The phenomenon of haemoglobin-releasing oxygen when the blood carbon dioxide concentration increases and taking up oxygen when the blood carbon dioxide levels decrease.

C

Carboxylation (HL)	The addition of carbon dioxide.
Carrier (SL)	An individual that has one copy of a recessive allele that causes a genetic disease in individuals that are homozygous for this allele.
Carrying capacity (HL-C)	The maximum population size supported by the environment in which the population lives.
Catabolism (SL)	The breaking down of large molecules into smaller molecules.
Cell respiration (SL)	The controlled release of energy from organic compounds in cells to form ATP.
Chemiosmosis (HL)	The movement of ions down the concentration gradient and across a semipermeable membrane, especially in the generation of ATP in cellular respiration and photosynthesis.
Chemoautotroph (HL)	An organism that uses energy from chemical reactions to generate ATP and produce organic compounds from inorganic substances.

Chemoheterotroph (HL)	An organism that uses energy from chemical reactions to generate ATP and obtain organic compounds from other organisms.
Chorionic villus sampling (SL)	A procedure whereby a small sample of chorionic villi is collected from the placenta of a pregnant woman and then tested to diagnose chromosomal abnormalities in the foetus.
Chromatography (SL)	A technique used in the separation of mixtures, especially in the separation of photosynthetic pigments.
Clade (SL)	A group of organisms consisting of all the descendants from a common ancestor.
Cladogram (SL)	A tree diagram that shows the most likely sequence of divergence in a clade.
Climograph (SL-C)	A graph representing the climate data, such as rainfall and temperature at a certain location.
Clone (SL)	A group of genetically identical organisms or a group of cells derived from a single parent cell.
Co-dominant alleles (SL)	Pairs of alleles that both affect the phenotype when present in a heterozygote.
Codon (SL)	The triplet of bases found on the mRNA.
Cohesion (SL)	The attraction of particles of the same type.
Community (SL)	A group of populations living and interacting with each other in an area.
Competitive exclusion (SL-C)	When more than one species occupies a niche, this results in competition for resources such as food or breeding sites, causing one species to be excluded from the ecosystem.
Competitive inhibition (HL)	A type of enzyme inhibition where the inhibitor binds to the active site and prevents the substrate from binding.
Consumer (SL)	An organism that ingests other organic matter that is living or recently killed.
Crossing over (SL)	A process occurring in prophase I of meiosis where non-sister chromatids of homologous chromosomes exchange segments of DNA so as to increase variation in offspring.
Cytokinesis (SL)	The splitting of the cytoplasm in a cell following mitosis.

D

Decarboxylation (HL)	The removal of carbon dioxide.
Denaturation (SL)	A structural change in a protein that results in the loss (usually permanent) of its biological properties.
Denitrification (HL-C)	The conversion of nitrate to nitrogen gas or nitrous oxide in the nitrogen cycle.
Depolarisation (SL)	The event resulting in the inside of the axon to become more positive than the outside, which allows for the transmission of an action potential.
Detritivore (SL)	An organism that ingests non-living organic matter.
Diastole (SL-D)	The part of the cardiac cycle where the chambers of the heart are relaxed and filling with blood.
Differentiation (SL)	Cells in multicellular organisms become differentiated or specialised in structure and function to perform a specific role for the organism.
Diffusion (SL)	The passive movement of particles from a region of high concentration to a region of low concentration.
Dihybrid (HL)	A cross between two individuals that differ in two traits that are being investigated.
Diploid (SL)	A cell containing twice the number of autosomal chromosomes (2n).
DNA profiling (SL)	A technique where an individual's DNA is mapped so that it can be compared to a reference sample. Useful in paternal testing and forensic investigations.
Domain (SL)	The highest level of taxonomy, where living organisms are divided into three groups, or domains.
Dominant allele (SL)	An allele that has the same effect on the phenotype whether it is present in the homozygous or heterozygous state.

E

Ecology (SL)	The study of relationships between living organisms and between organisms and their environment.
Ecosystem (SL)	A community and its abiotic environment.
Ectoderm (SL-A)	A layer in germ cell of the early embryo, which ultimately differentiates to form the brain, spine and peripheral nerves of the nervous system.
Embryo (SL)	A multicellular eukaryote in the first one to eight weeks following fertilisation.
Endocrine gland (SL-D)	Ductless glands that secrete hormones directly into the blood.
Endotoxins (HL)	Lipopolysaccharides in the walls of Gram-negative bacteria that cause fever and aches.
End-product inhibition (HL)	A form of negative feedback where the end product of one reaction becomes the inhibitor to the enzyme involved in the first reaction in order to regulate the amount of product produced.
Enzyme (SL)	A globular protein that acts as a catalyst of chemical reactions. Enzymes are specific for one type of reaction only.
Epidemiology (HL)	The study of the occurrence, distribution and control of diseases.
Ethology (HL-A)	The study of animal behaviour.
Evolution (SL)	The cumulative change in the heritable characteristics of a population.
Excretion (HL)	The removal from the body of the waste products of metabolic pathways.
Exocrine gland (SL-D)	Glands that secrete hormones into ducts.
Exotoxins (HL)	Specific proteins secreted by bacteria that cause symptoms such as muscle spasms (tetanus) and diarrhoea.
Exponential growth (HL-C)	Occurs when resources are unlimited in an environment and the population growth increase by a rapid rate in proportion to the growing total number of organisms.
Ex situ conservation (SL-C)	Where organisms are conserved away from their natural habitat such as in a zoo.

F

Fertilisation (HL)	The fusion of male and female gametes to form a new organism.
Fossil record (SL)	The record gained from the fossilisation of living organisms over time to show the occurrence and evolution of these organisms.

G

Gamete (SL)	The sex cells (sperm and egg), which fuse during fertilisation.
Gas exchange (SL)	The process of swapping one gas for another, which occurs at the alveoli and involves the swapping of gas from the air to the blood capillaries.
Gel electrophoresis (SL)	A technique used to separate DNA fragments based on their size and charge.
Gene (SL)	A heritable factor that controls a specific characteristic.
Gene mutation (SL)	A permanent change in the sequence of base pairs in the DNA that makes up a gene.
Gene pool (HL)	All of the genes of one species available in an interbreeding population.
Genome (SL)	The whole of the genetic information of an organism.
Genotype (SL)	The alleles of an organism.
Gersmehl diagram (SL-C)	A diagram showing the differences in nutrient levels between ecosystems.
Gram staining (SL-B)	A method used to differentiate between two groups of bacteria based on the properties of their cell walls which react differently when stained with violet dye.
Greenhouse gas (SL)	A gas present in Earth's atmosphere that absorbs and emits thermal radiation.

H

Habitat (SL)	The environment in which a species normally lives or the location of a living organism.
Haploid (SL)	A cell with a single set of unpaired chromosomes, or half the diploid number of a somatic cell.
Heterotroph (SL)	An organism that obtains organic molecules from other organisms.
Heterozygous (SL)	Having two different alleles of a gene.
Homologous chromosomes (SL)	Have the same genes as each other, in the same sequence, but not necessarily the same alleles of those genes, the same shape and size and the same banding pattern.
Homologous structure (SL)	Structures that have a common ancestry or origin but differ in their function.
Homologous trait (SL)	A characteristic derived from a common ancestor.
Homozygous (SL)	Having two identical alleles of a gene.
Human Genome Project (SL)	A research project undertaken internationally to determine the base sequence of human DNA and to map the human genome.
Hydrolysis (SL)	The splitting of water in a chemical reaction to form hydrogen ions and hydroxide ions.
Hydrophilic (SL)	Polar molecules that have an affinity for water and are often able to dissolve in water. Also termed 'water-loving'.
Hydrophobic (SL)	Non-polar molecules that tend to repel or not mix with water and therefore do not dissolve easily in water.

I

Immunity (SL)	The ability to resist infection or suffer symptoms of a particular disease due to the action of antibodies.
Imprinting (HL-A)	The tendency of a young animal to recognise another animal or person as a parent.
Independent assortment (HL)	The formation of random combinations of chromosomes from the mother and father during meiosis.
Indicator species (SL-C)	A species, whose presence or abundance in a particular environment, indicates certain ecological conditions within that environment.
Innate (HL-A)	Genetic/inherited, the same in each member of the same species.
Innate behaviour (SL)	Behaviour that develops independently of the environmental context.
In situ conservation (SL-C)	Where organisms are conserved within their natural habitat such as in a national park.

K

Karyogram (SL)	A photograph or diagram of all the chromosomes found within a cell. The chromosomes are arranged in their homologous pairs in order of size and are numbered.
Keystone species (SL-C)	A species that has a disproportionally large effect on the community in which it lives.
Knockout (HL-B)	A genetic technique where the target gene in an organism is rendered inoperative or 'knocked out' of the organism.

L

Learned behaviour (SL)	A behaviour that develops as a result of experience.
Limiting factor (SL)	The one factor that limits the rate of a reaction. In biology it is usually the factor that is furthest from its optimum value.

Linkage group (HL)	A group of genes whose loci are on the same chromosome.
Locus (SL)	The particular position on homologous chromosomes of a gene.

M

Malnutrition (SL-D)	A lack of nutrition caused by either not having enough to eat or not getting enough of the required nutrients from food that is eaten.
Meiosis (SL)	A type of cell division resulting in four daughter cells, with each having half the number of chromosomes as the parent cell (haploid). Used in the production of gametes in sexually reproducing organisms.
Meristem (HL)	A type of plant tissue found in the regions of the plant where growth occurs such as in the roots and shoots.
Metabolism (SL)	The sum of all reactions that occur in a cell or organism.
Microplastic (SL-C)	Small particles of plastic causing increasing problems in the environment, particularly marine environments.
Mitosis (SL)	A type of cell division resulting in two daughter cells, with each having the same number of chromosomes as the parent cell. Used when identical cells are needed.
Monoclonal antibodies (HL)	Antibodies that are produced by a single clone of cells and used in the laboratory.
Multicellular organism (SL)	An organism consisting of many cells.
Mutation (SL)	A change in a gene, which results in a variation and can be passed on to future offspring.
Myelination (SL)	A fatty layer called myelin covers the outside of neurons, which allows a nerve impulse to be passed along the neuron.
Myosin (HL)	A fibrous protein forming the contractile filaments of skeletal muscle and involved in muscle contraction.

N

Natural selection (SL)	A theory proposed by Charles Darwin whereby organisms better suited to their environments survive and reproduce, passing on these favourable traits to their offspring.
Negative feedback (SL)	The mechanism where a change is detected in the body and is then brought back to the norm.
Neural pruning (SL-A)	The process of eliminating synapses in the brain that are no longer useful during early childhood.
Neuron (SL)	A nerve cell responsible for transmitting electrical signals or nerve impulses throughout the nervous system.
Neurotransmitter (HL-A)	A chemical substance that diffuses across the synaptic cleft after being released from the end of a neuron following the arrival of a nerve impulse.
Niche (SL-C)	The way in which an organism fits into its community or ecosystem, which ultimately effects its survival as a species.
Nitrogen-fixing (HL-C)	The process where atmospheric nitrogen is incorporated into organic compounds in organisms.
Non-competitive inhibition (HL)	A type of enzyme inhibition where the inhibitor binds to a site away from the active site and prevents the substrate from binding.
Non-disjunction (SL)	The failure of a pair of homologous chromosomes or sister chromatids to separate during meiosis. Resulting in the daughter cell containing either three or one of the particular chromosomes.
Nucleosome (HL)	A length of DNA coiled around a group of eight histone proteins in eukaryotic cells.

O

Operant conditioning (HL-A)	A method of learning through a system of reward or punishment to alter the behaviour.
Osmosis (SL)	The passive movement of water molecules, across a partially permeable membrane, from a region of lower solute concentration to a region of higher solute concentration.
Osmoregulation (HL)	The control of the water balance of the blood, tissue or cytoplasm of a living organism.
Organ (SL)	A self-contained part of a multicellular organism with a specific and often essential function.
Organic compounds (SL)	Compounds containing carbon that are found in living organisms (except hydrogen carbonates, carbonates and oxides of carbon).
Osmolarity (SL)	The concentration of a solution in terms of its total number of solute particles per litre of solution.
Oxidation (HL)	A reaction where an element loses electrons and gains hydrogen ions due to the addition of oxygen.

P

Parasympathetic (HL-A)	A part of the autonomic nervous system responsible for the control of the state of rest in the body.
Partial pressure (HL)	The pressure exerted by a single component of a mixture of gases.
Passive immunity (HL)	Immunity due to the acquisition of antibodies from another organism in which active immunity has been stimulated, including via the placenta, colostrum, or by injection of antibodies.
Pathogen (SL)	An organism or virus that causes a disease.
PCR (SL)	Polymerase Chain Reaction, a method used in the laboratory to create multiple copies of the same segment of DNA.
Pedigree chart (SL)	A diagram representing the incidence of phenotypes for a particular gene within a family from one generation to the next.
Phenotype (SL)	The characteristics of an organism.
Phloem (HL)	A tissue in plants responsible for the translocation of sugars and other metabolic products from source (production) to sink (storage).
Phosphorylation (HL)	The addition of a phosphate group.
Photoautotroph (HL)	An organism that uses light energy to generate ATP and produce organic compounds from inorganic substances.
Photoheterotroph (HL)	An organism that uses light energy to generate ATP and obtains organic compounds from other organisms.
Photolysis (HL)	The separation of a molecule caused by the action of light energy.
Phylogenetics (HL-B)	The analysis of data to show the evolutionary development of a species or trait within a species.
Pneumocyte (HL-D)	A cell lining the alveoli in the lungs. May be a type I or type II pneumocyte.
Pollination (HL)	The transfer of pollen from the anther to the carpal must occur prior to fertilisation.
Polygenic inheritance (HL)	The inheritance of a characteristic which is controlled by more than one gene.
Population (SL)	A group of organisms of the same species that live in the same area at the same time.
Positive feedback (SL)	The increase in function as a result of responding to a stimulus.
Postsynaptic (HL-A)	Occurring after the synapse.
Presynaptic (HL-A)	Occurring before the synapse.
Primer (HL)	A strand of DNA or RNA required for the beginning of DNA replication.
Proteome (SL)	The complete set of proteins that can be expressed by a cell, tissue or organism.

Pyramid of energy (SL)	A model showing the flow of energy in an ecosystem from the producers through to the top-order consumers.

R

Random orientation (SL)	The random orientation of homologous chromosomes along the metaphase plate during Metaphase I of meiosis.
Reabsorption (HL)	In the kidneys, the flow of glomerular filtrate from the proximal convoluted tubule into the surrounding capillaries.
Receptor (SL-A)	Part of an organ, tissue or cell that is able to respond to external stimuli such as light.
Recessive allele (SL)	An allele that only has an effect on the phenotype when present in the homozygous state.
Recombinant (HL)	An organism containing a different combination of alleles than either the father or the mother.
Recombinant DNA (SL-B)	Artificially formed DNA using segments of DNA from different organisms and splicing them together.
Reduction (HL)	A reaction where an element gains electrons and loses hydrogen ions. Reduction is the opposite reaction to oxidation and the two normally occur together.
Reflex (HL-A)	A rapid, unconscious response.
Reflex arc (HL-A)	The nerve pathway involved in a reflex action, beginning at the receptor, travelling to the spinal cord and then to the effector to initiate the response.
Repolarisation (SL)	The change in potential across the membrane of an axon, returning it to a negative value following the passing of an action potential.
Response (SL)	An action resulting from the perception of a stimulus. A reaction to change that is perceived by the nervous system.
Resting potential (SL)	An electrical impulse across a cell membrane when not propagating an impulse.

S

Saprotroph (SL)	An organism that lives on or in non-living organic matter, secreting digestive enzymes into it and absorbing the products of digestion.
Sarcomere (HL)	The contractile unit of skeletal muscle.
Seed dispersal (HL)	The moving of seeds away from the parent plant to reduce competition.
Semi-conservative (SL)	The nature of DNA replication, where the newly replicated DNA is comprised of one old strand and one new strand.
Sense strand (HL)	Coding strand that has the same base sequence as mRNA with uracil instead of thymine.
Sex chromosomes (SL)	Are those which help determine the sex of an individual.
Sex linkage (SL)	Genes which are carried on the X or Y chromosome.
Sinusoid (SL-D)	A specialised capillary found in the liver.
Sister chromatid (SL)	One half of a duplicated chromosome.
Speciation (HL)	The evolutionary process resulting in the formation of new species.
Species (SL)	A group of organisms that can interbreed and produce fertile offspring.
Standard Deviation (SD) (SL)	Used to summarise the spread of values around the mean, with 68 per cent of the values falling within one SD of the mean.
Stem cells (SL)	Undifferentiated cells capable of differentiating into other kinds types of cells in a multicellular organism.
Stimulus (SL)	A change in the environment (internal or external) that is detected by a receptor and elicits a response.
Substrate (SL)	The substance upon which an enzyme acts.

Supercoil (SL)	The nature of the DNA double helix having undergone additional twisting in order to allow it to fit inside the nucleus.
Symbiotic (SL-C)	A relationship between two organisms of different species, which may be beneficial or harmful.
Sympathetic (HL-A)	Part of the autonomic nervous system responsible for the activation of the fight or flight response in stressful or dangerous situations.
Synapse (SL)	The junction or gap between two neurons.
Systole (SL-D)	The part of the cardiac cycle where the heart muscle contracts and pumps blood.

T

Test cross (SL)	Testing a suspected heterozygote by crossing it with a known homozygous recessive.
Threshold potential (SL)	The level at which the potential across the membrane of an axon becomes sufficiently less negative to initiate an action potential.
Tissue (SL)	A collection of specialised cells within a multicellular organism.
Transect (SL-C)	A sampling technique using a line along which the number of individuals of a species are counted.
Transgenic organisms (SL-B)	An organism with an altered genome.
Transpiration (HL)	The loss of water vapour from the leaves and stems of plants.
Trophic level (SL)	The level of the food chain at which an organism is found.

U

Ultrafiltration (HL)	A process occurring in the Bowman's capsule of the kidney where urea, salts, water and glucose are removed from the blood to form the glomerular filtrate.
Ultrastructure (SL)	The structures within a cell visible with an electron microscope.

V

Variation (SL)	A difference between cells or organisms caused by the effect of environmental factors, genetic mutation or arising from sexual reproduction.
Ventilation (SL)	The exchange in the lung of stale air with fresh air.
Viral vector (HL-B)	A virus is used in a technique to deliver genetic material into cells.

W

Wavelength (SL)	The measure of the length of the waves in light energy in nanometres.

X

Xylem (HL)	A plant tissue involved in the transport of water from the roots of the plant to the stems and leaves.

Z

Zygote (SL)	A fertilised egg or ovum that results from the fusion of sperm and egg.

Appendix 3
EXAM PREPARATION

When studying for the examination:
- Have the course outline for the topic in front of you so you can check on any requirements.
- Take short but frequent breaks. This rests your mind and helps you focus more easily.
- Find a study technique that suits you, or use a range of techniques to ensure you are ready.
- If you are having trouble keeping motivated, set yourself a time limit and a goal for work you want to achieve before that time.
- Use past examination papers as a guide to the style of questions to expect.
- Test your knowledge with practice activities and questions.
- Discover your preferred learning style and tailor your study to suit this style.

Before the examination:
- Get lots of rest and sleep. Your studying is of no use if you cannot stay awake during the exam.
- Ensure you have something healthy to eat and go to the toilet.
- Go for a walk and get some fresh air or sit quietly to relax yourself.
- Take a bottle of water with you into the examination room, but don't drink too much.

During the examination:
- Do the questions that you know first. This allows you to feel like you're getting somewhere.
- Remember to look at the marks allocated to each question. This gives you a guide as to how long the question should take you and how many points or ideas you need to include.
- For multiple-choice questions there is only one correct answer; however, there may be a situation where two are correct, but one is more correct than the other. Choose carefully.
- When answering a multiple-choice question that you don't automatically know, go through each of the answers and cross off the ones that you know are definitely wrong. You may write notes next to these to help you out. If there are still questions you're not sure of, leave them and move on. But remember to go back, never leave a multiple-choice question unanswered.
- Ensure you read the questions carefully; there is no point spending 10 minutes writing a perfect answer to a question that wasn't even asked.
- Make the most of reading time to read through the questions, check your paper has all the pages and that it is for the subject you are studying.

Revision Techniques

Examination revision is most effective when the student uses techniques most suited to their preferred learning style. The suggestions below are categorised based on the four main learning styles.

Visual learners

- Draw concept maps to visually organise information in large topics
- Create flow charts of important processes.
- Draw summary diagrams of complex processes.
- Use colour to enhance understanding and recall.
- Use the internet to search for video clips relating to the current topic of study.
- Draw, colour, label or annotate diagrams.

Auditory learners

- Read notes aloud and record yourself. This can be played back on your iPod whilst on the bus or going for a walk.
- Use the internet to search for videos or audio recordings relating to the current topic of study.
- Explain an important concept aloud to a parent or friend.
- Organise a study group with other auditory learners to discuss current topics.
- Ask a parent to test you aloud with prepared questions and answers.

Read/write learners

- Summarise your text or notes. Then reduce the summary even further until you have condensed the notes into more manageable chunks.
- Use highlighters to
  - Write out ordered steps for a challenging or complex process
  - Organise important information into lists
  - Annotate a diagram with relevant extra details
  - Compile a glossary of key terms and their definitions.

Kinaesthetic learners

- Write yourself some practice questions and attempt these.
- Summarise the information and convert it into a different form. For example, convert large sections of writing into a summary table.
- Ask a parent to test you aloud with prepared questions and answers.
- Write out the steps to a complex process on separate pieces of paper, jumble them up and then reorder them correctly.

Constructing Extended Response Answers

Extended response questions from Section B of Paper 2. Each set of extended response questions is allocated 16 marks. Fifteen of these marks are for content and one mark is for the quality of the response.

The following are provided as suggestions to gain the mark for quality:

- Answer in full sentences, do not simply list points as dot points.
- Use paragraphs and separate your answer.
- Begin each extended response with a definition of any key terms.
- Take note of the command term used in the question as this will give a good indication of the depth of answer required.
- Ensure to read the question carefully.
- Use scrap paper to plan your response.

- Answer all parts to the question.
- Read over your response when finished to ensure it flows and can be clearly understood.
- Never cross out any part of your answer, unless it contradicts what you would like to say.
- Write neatly and legibly.
- Use the marks allocated to each question as a guide to the number of points required.

References

Adelaide Microscopy, 2014, *Cillia in Protozoan (TEM)*, University of Adelaide, Australia.

Adelaide Microscopy, 2014, *Dunaliella salina (TEM)*, University of Adelaide, Australia.

Adelaide Microscopy, 2014, *Mitochondria (TEM)*, University of Adelaide, Australia.

Bockman, P., 2011, *Primate Cladogram*, available from http://commons.wikimedia.org/wiki/File:Primate_cladogram.jpg.

Glenlarson, 2007, *Lead Generated Sinus Rhythm*, available from http://commons.wikimedia.org/wiki/File%3A12_lead_generated_sinus_rhythm.JPG.

Gould, J., 1804–81, *Galapagos finches* (public domain), available from http://commons.wikimedia.org/wiki/File%3ADarwin's_finches.jpeg.

International Baccalaureate Organization, 2014, *Diploma Programme Biology Guide*, Cardiff, United Kingdom, pp. 166–67.

Jacobson, G., 1915, *Beetles Russia and Western Europe* (public domain), available from http://commons.wikimedia.org/wiki/Category:Georgiy_Jacobson._Beetles_Russia_and_Western_Europe_(extracted_images).

Plant and Soil Sciences eLibrary, 2015, 'Chi-Square Distribution Table', University of Nebraska-Lincoln, USA, available from http://passel.unl.edu/pages/informationmodule.php?idinformationmodule=1130447119.

Roschier, L., 2009, *Body Mass Index Nomogram*, available from http://pynomo.org/wiki/index.php?title=Body-mass_index.

Sasol Inzalo Foundation, 2012, *Natural Sciences and Technology Grade 5*, available from http://www.thunderboltkids.co.za/Grade5/01-life-and-living/chapter2.html.

Index

Milton Keynes UK
Ingram Content Group UK Ltd.
UKHW050703290124
436892UK00016B/514